FastAPI Cookbook

Develop high-performance APIs and web applications
with Python

Giunio De Luca

FastAPI Cookbook

Group Product Manager: Kaustubh Manglurkar

Publishing Product Manager: Bhavya Rao

Book Project Manager: Arul Viveaun S

Senior Editor: Nathanya Dias

Technical Editor: Simran Ali

Copy Editor: Safis Editing

Indexer: Manju Arasan

Production Designer: Jyoti Kadam

DevRel Marketing Coordinators: Anamika Singh and Nivedita Pandey

First published: August 2024

Production reference: 2040924

Published by Packt Publishing Ltd.
Grosvenor House
11 St Paul's Square
Birmingham
B3 1RB, UK

ISBN 978-1-80512-785-7

www.packtpub.com

To my dear nephew, Marco.

You bring endless joy to our lives. I wish you a future filled with love, growth, and happiness. May you always find success and fulfillment in everything you do.

– Giunio De Luca

Foreword

Having worked alongside Giunio for over a year, I can confidently say that his expertise and passion for Python development are evident in every line of code he produces. I affirm this with no shadow of a doubt, as I am a developer with over 30 years of experience across several countries and prominent industries, including Microsoft, NATO, and IBM.

We naturally became friends due to our mutual respect and shared interests, both professional and personal. Giunio stood out in our team at Coreso, one of the European Regional Coordination Centers for high-voltage electricity flows. He has an impressive academic background coupled with extensive international experience. What truly sets him apart, however, is his deep and thorough understanding of Python and FastAPI. Giunio's code is not only functional but also exceptionally clean and clear—a hallmark of a skilled developer, or in other words, a *subject matter expert*.

Beyond his technical prowess, Giunio possesses a quality that I find equally important: an open mind. He readily embraces new challenges, tackling them with both enthusiasm and a solid foundation of knowledge, which he diligently builds and extends if he does not already possess it. I believe that he has brought these qualities to fruition in creating this book, pushing it beyond the scope of a typical cookbook. For instance, dealing with real-time communication and WebSockets is something rarely found in other cookbooks, and finding the right solution for your needs can consume a lot of valuable time searching online.

This book promises to be a valuable resource for anyone seeking to become proficient in the use of FastAPI. Despite the excellent work the creators do with the product documentation, navigating it can easily lead to hours of frustration. Whether you're a seasoned Python developer or just starting your journey, Giunio's clear explanations and practical examples will guide you every step of the way.

So, let Giunio be your expert guide to the power and efficiency of FastAPI!

Antonio Ferraro

MSc Computer Science, Business Owner

Contributors

About the author

Giunio De Luca is a software engineer with over 10 years of experience in fields such as physics, sports, and administration. He graduated in industrial engineering from the University of Basilicata and holds a PhD in numerical simulations from Paris-Saclay University. His work spans developing advanced algorithms, creating sports analytics applications, and improving administrative processes. As an independent consultant, he collaborates with research labs, government agencies, and start-ups across Europe. He also supports coding education in schools and universities through workshops, lectures, and mentorship programs, inspiring the next generation of software engineers with his expertise and dedication.

With heartfelt gratitude, I thank my family and my loved ones for their continuous encouragement. I am also thankful to all my employers who have believed in my expertise, and my current clients and colleagues who keep trusting my work and providing me with valuable opportunities for professional growth.

About the reviewer

Adarsh Divakaran is an experienced backend developer with expertise in architecting, building, and deploying APIs, primarily using Python. He is the co-founder of Digievo Labs, a global technology firm. He loves reading all sorts of Python content and runs a newsletter, *Python in the Wild*, showcasing Python concepts and examples from open source projects. Adarsh also speaks at global Python conferences, including PyCascades, EuroPython, and FlaskCon.

Table of Contents

viii Table of Contents

3

Building RESTful APIs with FastAPI 51

4

Authentication and Authorization   77

5

Testing and Debugging FastAPI Applications   109

6

Integrating FastAPI with SQL Databases   135

7

Integrating FastAPI with NoSQL Databases   167

8

Advanced Features and Best Practices   201

9

Working with WebSocket   225

10

Integrating FastAPI with other Python Libraries

11

Middleware and Webhooks

12

Deploying and Managing FastAPI Applications   301

Preface

FastAPI Cookbook is a vital resource for Python developers who want to master the FastAPI framework to build APIs. Created by Sebastián Ramírez Montaño, FastAPI was first released in December 2018. It quickly gained popularity and became one of the most widely used Python frameworks for building APIs.

The book starts by introducing FastAPI, showing its advantages, and it will help you set up your development environment. It then moves on to data handling, showing database integration and **create, read, update and delete (CRUD)** operations, to help you manage data effectively within your APIs.

As the book progresses, it explores how to create **RESTful** APIs, covering advanced topics such as complex queries, versioning, and extensive documentation. Security is also important, and the book has a chapter on implementing authentication mechanisms such as **OAuth2** and **JWT** tokens to secure FastAPI applications.

Testing is an essential part of development, and the book offers strategies to ensure the quality and reliability of FastAPI applications. Deployment strategies are discussed, highlighting best practices for production environments. For applications with high traffic, the book examines scaling techniques to improve performance.

Extending FastAPI's functionality is possible through middleware, and the book also demonstrates how to boost FastAPI's capabilities by integrating it with other Python tools and frameworks to accommodate machine learning models and expose **LLM RAG** applications.

Real-time communication is handled with a chapter on **WebSockets**, and advanced data handling techniques are provided to manage large datasets and file management.

The book ends with a focus on serving real-world traffic with FastAPI, stressing deployment strategies and packaging shipping. Each chapter is carefully designed to build your expertise, making the *FastAPI Cookbook* a valuable guide for professional-grade API development.

Who this book is for

The book is tailored for intermediate to advanced Python developers who have a foundational understanding of web development concepts. It's particularly beneficial for those seeking to build efficient, scalable APIs with the modern FastAPI framework. The book is a valuable resource for developers looking to enhance their API development skills and apply practical solutions to real-world programming challenges. Whether you're looking to secure APIs, manage data effectively, or optimize performance, this book provides the knowledge and hands-on examples to elevate your expertise in FastAPI.

What this book covers

Chapter 1, First Steps with FastAPI, serves as an introduction to the framework, emphasizing its speed, ease of use, and comprehensive documentation. This chapter is the gateway for you to set up your development environment, create your first FastAPI project, and explore its fundamental concepts.

Chapter 2, Working with Data, is dedicated to mastering the critical aspect of data handling in web applications. It covers the intricacies of integrating, managing, and optimizing data storage using both SQL and NoSQL databases.

Chapter 3, Building RESTful APIs with FastAPI, dives into the essentials of constructing RESTful APIs, which are fundamental to web services, enabling applications to communicate and exchange data efficiently.

Chapter 4, Authentication and Authorization, delves into the critical realms of securing your web applications against unauthorized access. It covers the basics of user registration and authentication, the integration of OAuth2 protocols with JWT for enhanced security, and the creation of essential components for an API.

Chapter 5, Testing and Debugging FastAPI Applications, pivots toward a crucial aspect of software development that ensures the reliability, robustness, and quality of your applications – testing and debugging.

Chapter 6, Integrating FastAPI with SQL Databases, embarks on a journey to harness the full potential of SQL databases within FastAPI applications. It is meticulously designed to guide you through leveraging SQLAlchemy, a powerful SQL toolkit and **Object-Relational Mapper (ORM)** for Python.

Chapter 7, Integrating FastAPI with NoSQL Databases, explores the integration of FastAPI with NoSQL databases by guiding you through the process of setting up and using MongoDB, a popular NoSQL database, with FastAPI. It covers CRUD operations, working with indexes for performance optimization, and handling relationships in NoSQL databases. Additionally, the chapter discusses integrating FastAPI with Elasticsearch for powerful search capabilities and implementing caching using Redis.

Chapter 8, Advanced Features and Best Practices, explores advanced techniques and best practices to optimize the functionality, performance, and scalability of FastAPI applications. It covers essential topics such as dependency injection, custom middleware, internationalization, performance optimization, rate limiting, and background task execution.

Chapter 9, Working with WebSockets, is a comprehensive guide to implementing real-time communication features in FastAPI applications using WebSockets. It covers setting up WebSocket connections, sending and receiving messages, handling connections and disconnections, error handling, and implementing chat functionality.

Chapter 10, Integrating FastAPI with other Python Libraries, deep dives into the potential of FastAPI when coupled with external libraries, enhancing its capabilities beyond its core features. It provides a recipe-based approach to integrating FastAPI with various technologies, such as Cohere and LangChain, to build LLM RAG applications.

Chapter 11, Middleware and Webhooks, delves into the advanced and crucial aspects of middleware and Webhooks in FastAPI. Middleware allows you to process requests and responses globally, while Webhooks enable your FastAPI application to communicate with other services by sending real-time data updates.

Chapter 12, Deploying and Managing FastAPI Applications, covers the knowledge and tools needed to deploy FastAPI applications seamlessly, leveraging various technologies and best practices. You will learn how to utilize the FastAPI CLI to run your server efficiently, enable HTTPS to secure your applications, and containerize your FastAPI projects with Docker.

To get the most out of this book

You should have a fundamental understanding of Python programming, as the book assumes familiarity with Python syntax and concepts. Additionally, knowledge of web development principles, including HTTP, RESTful APIs, and JSON, will be beneficial. Familiarity with SQL and NoSQL databases, as well as experience with version control systems such as Git, will help you to fully grasp the content.

Software/hardware covered in the book	OS requirements
Python 3.9 or higher	Windows, macOS, or Linux (any)

If you are using the digital version of this book, we advise you to type the code yourself or access the code via the GitHub repository (link available in the next section). Doing so will help you avoid any potential errors related to the copying and pasting of code.

Download the example code files

You can download the example code files for this book from GitHub at `https://github.com/PacktPublishing/FastAPI-Cookbook`. If there's an update to the code, it will be updated on the existing GitHub repository.

We also have other code bundles from our rich catalog of books and videos available at `https://github.com/PacktPublishing/`. Check them out!

Conventions used

There are a number of text conventions used throughout this book.

`Code in text`: Indicates code words in text, database table names, folder names, filenames, file extensions, pathnames, dummy URLs, user input, and Twitter handles. Here is an example: "Also, you will find only the messages from our `logger_client` in a newly created `app.log` file automatically created by the application."

A block of code is set as follows:

```
from locust import HttpUser, task


class ProtoappUser(HttpUser):
    host = "http://localhost:8000"

    @task
    def hello_world(self):
        self.client.get("/home")
```

When we wish to draw your attention to a particular part of a code block, the relevant lines or items are set in bold:

```
from pydantic import BaseModel, Field

class Book(BaseModel):
    title: str = Field(..., min_length=1, max_length=100)
    author: str = Field(..., min_length=1, max_length=50)
    year: int = Field(..., gt=1900, lt=2100)
```

Any command-line input or output is written as follows:

```
$ pytest --cov protoapp tests
```

Throughout this book, we will generally use Unix-like terminal commands. This might lead to compatibility issues with Windows for commands that run on multiple lines. If you are using a Windows terminal, consider adapting the newline character \ as follows:

```
$ python -m grpc_tools.protoc \
--proto_path=. ./grpcserver.proto \
--python_out=. \
--grpc_python_out=.
```

Here is the same line in CMD:

```
$ python -m grpc_tools.protoc ^
--proto_path=. ./grpcserver.proto ^
--python_out=. ^
--grpc_python_out=.
```

Here is the line in Powershell:

```
$ python -m grpc_tools.protoc `
--proto_path=. ./grpcserver.proto `
--python_out=. `
--grpc_python_out=.
```

Bold: Indicates a new term, an important word, or words that you see onscreen. For example, words in menus or dialog boxes appear in the text like this. Here is an example: "This limit can be adjusted in the settings (**Settings | Advanced Settings | Run/Debug | Temporary configurations limit**)."

> **Tips or important notes**
> Appear like this.

Sections

In this book, you will find several headings that appear frequently (*Getting ready*, *How to do it...*, *How it works...*, *There's more...*, and *See also*).

To give clear instructions on how to complete a recipe, use these sections as follows.

Getting ready

This section tells you what to expect in the recipe and describes how to set up any software or preliminary settings required for it.

How to do it...

This section contains the steps required to follow the recipe.

How it works...

This section usually consists of a detailed explanation of what happened in the previous section.

There's more...

This section consists of additional information about the recipe in order to make you more knowledgeable about the recipe.

See also

This section provides helpful links to other useful information for the recipe.

Get in touch

Feedback from our readers is always welcome.

General feedback: If you have questions about any aspect of this book, mention the book title in the subject of your message and email us at customercare@packtpub.com.

Errata: Although we have taken every care to ensure the accuracy of our content, mistakes do happen. If you have found a mistake in this book, we would be grateful if you would report this to us. Please visit www.packtpub.com/support/errata, select your book, click on the **Errata Submission Form** link, and enter the details.

Piracy: If you come across any illegal copies of our works in any form on the internet, we would be grateful if you would provide us with the location address or website name. Please contact us at copyright@packt.com with a link to the material.

If you are interested in becoming an author: If there is a topic that you have expertise in and you are interested in either writing or contributing to a book, please visit authors.packtpub.com.

Share Your Thoughts

Once you've read *FastAPI Cookbook*, we'd love to hear your thoughts! Scan the QR code below to go straight to the Amazon review page for this book and share your feedback.

https://packt.link/r/1-805-12785-3

Your review is important to us and the tech community and will help us make sure we're delivering excellent quality content.

Download a free PDF copy of this book

Thanks for purchasing this book!

Do you like to read on the go but are unable to carry your print books everywhere?

Is your eBook purchase not compatible with the device of your choice?

Don't worry, now with every Packt book you get a DRM-free PDF version of that book at no cost.

Read anywhere, any place, on any device. Search, copy, and paste code from your favorite technical books directly into your application.

The perks don't stop there, you can get exclusive access to discounts, newsletters, and great free content in your inbox daily

Follow these simple steps to get the benefits:

1. Scan the QR code or visit the link below

https://packt.link/free-ebook/978-1-80512-785-7

2. Submit your proof of purchase
3. That's it! We'll send your free PDF and other benefits to your email directly

1

First Steps with FastAPI

Welcome to the exciting world of **FastAPI**, a modern, high-performance framework for building APIs and web applications in **Python**. This first chapter is your gateway to understanding and harnessing the power of FastAPI. Here, you'll take your initial steps into setting up your development environment, creating your very first FastAPI project, and exploring its fundamental concepts.

FastAPI stands out for its speed, ease of use, and comprehensive documentation, making it a preferred choice for developers looking to build scalable and efficient web applications. In this chapter, you'll practically engage in setting up FastAPI, learning how to navigate its architecture, and understanding its core components. You'll gain hands-on experience by defining simple API endpoints, handling HTTP methods, and learning about request and response handling. These foundational skills are crucial for any developer stepping into the world of modern web development with FastAPI.

By the end of this chapter, you will have a solid understanding of FastAPI's basic structure and capabilities. You'll be able to set up a new project, define API endpoints, and have a grasp on handling data with FastAPI. This knowledge sets the stage for more advanced topics and complex applications you'll encounter as you progress through the book.

In this chapter, we're going to cover the following recipes:

- Setting up your development environment
- Creating a new FastAPI project
- Understanding FastAPI basics
- Defining your first API endpoint
- Working with path and query parameters
- Defining and using request and response models
- Handling errors and exceptions

Each recipe is designed to provide you with practical knowledge and direct experience, ensuring that by the end of this chapter, you'll be well equipped to start building your own FastAPI applications.

Technical requirements

To embark on your journey with FastAPI, you'll need to set up an environment that supports Python development and FastAPI's functionalities. Here's a list of the technical requirements and installations needed for this chapter:

- **Python**: FastAPI is built on Python, so you'll need a Python version compatible with your FastAPI version. You can download the latest version of it from `python.org`.

- **FastAPI**: Install FastAPI using `pip`, Python's package manager. You can do it by running `pip install fastapi` from the command terminal.

- **Uvicorn**: FastAPI requires an **Asynchronous Server Gateway Interface** (**ASGI**) server, and Uvicorn is a lightning-fast ASGI server implementation. Install it using `pip install uvicorn`.

- **Integrated development environment (IDE)**: An IDE such as **Visual Studio Code** (**VS Code**), PyCharm, or any other IDE that supports Python development will be necessary for writing and testing your code.

- **Postman or Swagger UI**: For testing API endpoints. FastAPI automatically generates and hosts Swagger UI, so you can use it right out of the box.

- **Git**: Version control is essential, and Git is a widely used system. If not already installed, you can get it from `git-scm.com`.

- **GitHub account**: A GitHub account is required to access the code repositories. Sign up at `github.com` if you haven't already.

The code used in the chapter is available on GitHub at the following address: `https://github.com/PacktPublishing/FastAPI-Cookbook/tree/main/Chapter01`. You can clone or download the repository at `https://github.com/PacktPublishing/FastAPI-Cookbook` to follow along on your local machine.

Setting up your development environment

This recipe, dedicated to setting up your development environment, is a critical foundation for any successful project in web development. Here, you'll learn how to install and configure all the essential tools needed to start building with FastAPI.

We begin by guiding you through the installation of Python, the core language behind FastAPI. Next, we'll move on to installing FastAPI itself, along with Uvicorn, a lightning-fast ASGI server, which serves as the bedrock for running your FastAPI applications.

Setting up an IDE is our next stop. Whether you prefer VS Code, PyCharm, or any other Python-friendly IDE, we'll provide tips to make your development process smoother and more efficient.

Lastly, we'll introduce you to Git and GitHub – indispensable tools for version control and collaboration in modern software development. Understanding how to use these tools will not only help you manage your code effectively but also open doors to the vast world of community-driven development and resources.

Getting ready

FastAPI works with Python, so you need to check your Python version before using it. This is an important step for setting up FastAPI. We will guide you through how to install it.

Windows installation

If you work on Windows, follow these steps to install Python:

1. Visit the official Python website: `python.org`.
2. Download the latest version of Python or any other version higher than 3.9.
3. Run the installer. Ensure to check the box that says **Add Python to PATH** before clicking **Install Now**.
4. After the installation, open Command Prompt and type `python --version` to confirm the installation.

macOS/Linux installation

macOS usually comes with Python pre-installed; however, it might not be the latest version.

You can use `Homebrew` (a package manager for macOS). To install it, open the terminal and run it:

```
$ /bin/bash -c "$(curl -fsSL https://raw.githubusercontent.com/\
Homebrew/install/HEAD/install.sh)"
```

Then, you can install Python – still from the terminal – with the following command:

```
$ brew install python
```

On Linux, you can install Python using the package manager by running the following command:

```
$ sudo apt-get install python3
```

That's all you need to install Python on macOS and Linux systems.

Checking the installation

You can then check that Python is correctly installed by running the following command in the terminal:

```
$ python --version
```

If you installed it on Linux, the binary command is python3, so you can check that Python is correctly installed by running the following command:

```
$ python3 --version
```

Once Python is installed, we want to make sure that the Python's package manager is correctly installed. It comes with Python's installation, and it's called pip.

From a terminal window, run the following command:

```
$ pip --version
```

On Linux, run the following command:

```
$ pip3 --version
```

Once Python is installed on your computer, you can now consider installing FastAPI.

How to do it...

When you have Python and pip ready, we can continue with installing FastAPI, the IDE. Then, we will configure Git.

We will do it by following these steps:

1. Installing FastAPI and Uvicorn
2. Setting up your IDE (VS Code or PyCharm)
3. Setting up Git and GitHub to track your project

Installing FastAPI and Uvicorn

With Python set up, the next step is installing FastAPI and Uvicorn. FastAPI is the framework we'll use to build our applications, and Uvicorn is an ASGI server that runs and serves our FastAPI applications.

Open your command-line interface and install FastAPI and Uvicorn together by running the following command:

```
$ pip install fastapi[all]
```

This command installs FastAPI along with its recommended dependencies, including Uvicorn.

To verify the installation, you can simply run uvicorn --version from the terminal.

Setting up your IDE

Choosing the right IDE is a crucial step in your FastAPI journey. An IDE is more than just a text editor; it's a space where you write, debug, and test your code.

A good IDE can significantly enhance your coding experience and productivity. For FastAPI development and Python in general, two popular choices are VS Code and PyCharm.

VS Code

VS Code is a free, open source, lightweight IDE with powerful features. It offers excellent Python support and is highly customizable.

You can download and install VS Code from the official website (`code.visualstudio.com`). The installation is quite straightforward. Once installed, open VS Code, go to **Extensions** (a square icon on the left bar), and search for `python`. Install the Microsoft version, and that is it.

PyCharm

PyCharm, created by JetBrains, is specifically tailored for Python development. It offers a broad range of tools for professional developers, including excellent support for web development frameworks such as FastAPI.

You can choose between a Community free edition and a Professional paid version. For the scope of the book, the Community Edition is largely sufficient, and it can be downloaded on the JetBrains website: `https://www.jetbrains.com/pycharm/download/`.

For PyCharm as well, the installation is straightforward.

Enhancing your development experience

For both IDEs – and if you use another of your choice – make sure to leverage basic perks to improve your experience as a developer and be more efficient. Here is a short checklist that I use when I approach a new IDE environment:

- **Code completion and analysis**: Good IDEs provide intelligent code completion, error highlighting, and fixes, which are invaluable for efficient development
- **Debugging tools**: Utilize debugging features provided by the IDE to diagnose and resolve issues in your code
- **Version control integration**: A good IDE offers support for Git, simplifying code change tracking and repository management
- **Customization**: Customize your IDE by adjusting themes, key binding, and settings to match your workflow, making your development experience as comfortable and productive as possible

Setting up Git and GitHub

Version control is an essential aspect of software development. Git, coupled with GitHub, forms a powerful toolset for tracking changes, collaborating, and maintaining the history of your projects. You can download the Git installer from the official website `git-scm.com` and install it.

After installation, configure Git with your username and email using the following commands in the command line:

```
$ git config --global user.name "Your Name"
$ git config --global user.email "your.email@example.com"
```

GitHub is the platform chosen to store code examples used in the book. Sign up for a GitHub account at github.com if you don't already have one.

Creating a new FastAPI project

Setting up a well-organized project structure is crucial for maintaining a clean code base, especially as your application grows and evolves. This recipe will guide you on how to create your first basic FastAPI project. A structured project simplifies navigation, debugging, and collaboration. For FastAPI, following best practices in structuring can significantly enhance scalability and maintainability.

Getting ready

All you need to do to follow the recipe is make sure that you have your development environment set up.

How to do it...

We begin by making a project folder named fastapi_start that we'll use as the root project folder.

1. From the terminal at the root project folder level, we'll set up our virtual environment by running the following command:

    ```
    $ python -m venv .venv
    ```

    This will create a .venv folder that will contain all packages required for the project within our project's root folder.

2. Now, you need to activate the environment. If you are on Mac or Linux, run the following command:

    ```
    $ source .venv/bin/activate
    ```

    From Windows, run the following command:

    ```
    $ .venv\Scripts\activate
    ```

    When the environment is active, you should see in your terminal a prefix string such as (.venv) $. Alternatively, if you check the location of the python binary command, it should be located within the .venv folder. From now on, each time you install a module with pip, it will be installed in the .venv folder, and it will be activated only if the environment is active.

3. Now, you can install the `fastapi` package with `uvicorn` in your environment by running the following command:

    ```
    $ pip install fastapi uvicorn
    ```

    Once FastAPI is installed in your environment, open your project folder with your favorite IDE and create a file called `main.py`.

4. This file is where your FastAPI application begins. Start by writing the import of the `FastAPI` module. Then, create an instance of the `FastAPI` class:

    ```
    from fastapi import FastAPI

    app = FastAPI()
    ```

    This instance houses the code of your application.

5. Next, define your first route. Routes in FastAPI are like signposts that direct requests to the appropriate function. Start with a simple route that returns a greeting to the world:

    ```
    @app.get("/")
    def read_root():
        return {"Hello": "World"}
    ```

    You've just created the code for your first FastAPI application.

If you want to track the project, you can set up Git as follows:

1. In your project's root directory, open a terminal or Command Prompt and run the following command:

    ```
    $ git init
    ```

    This simple command prepares your project for version control under Git.

    Before committing, create a `.gitignore` file to specify untracked files to ignore (such as `__pychache__`, .venv, or IDE-specific folders). You can also have a look at the one on the GitHub repository of the project at the link: `https://github.com/PacktPublishing/FastAPI-Cookbook/blob/main/.gitignore`.

2. Then, add your files with the following command:

    ```
    $ git add .
    ```

3. Then, commit them using the following command:

    ```
    $ git commit -m "Initial commit"
    ```

And that's it. You are now tracking your project with Git.

There's more...

A well-structured project is not just about neatness; it's about creating a sustainable and scalable environment where your application can grow and evolve. In FastAPI, this means organizing your project in a way that separates different aspects of your application logically and efficiently.

There is no unique and perfect structure for a FastAPI project; however, a common approach is to divide your project into several key directories:

- `/src`: This is where your primary application code lives. Inside `/src`, you might have subdirectories for different modules of your application. For instance, you could have a `models` directory for your database models, a `routes` directory for your FastAPI routes, and a `services` directory for business logic.
- `/tests`: Keeping your tests separate from your application code is a good practice. It makes it easier to manage them and ensures that your production builds don't include test code.
- `/docs`: Documentation is crucial for any project. Whether it's API documentation, installation guides, or usage instructions, having a dedicated directory for documentation helps maintain clarity.

See also

You can find detailed information on how to manage virtual environments with venv at the following link:

- *Creation of virtual environments*: `https://docs.python.org/3/library/venv.html`

To brush up your knowledge with Git and get familiar with adding, staging and commiting operations, have a look at this guide:

- *Git simple guide*: `https://rogerdudler.github.io/git-guide/`

Understanding FastAPI basics

As we embark on our journey with FastAPI, it's essential to build a solid foundation. FastAPI isn't just another web framework; it's a powerful tool designed to make your life as a developer easier, your applications faster, and your code more robust and maintainable.

In this recipe, we'll demystify the core concepts of FastAPI, delve into its unique features such as asynchronous programming, and guide you through creating and organizing your first endpoints. By the end of the recipe, you'll have your first FastAPI server up and running – a milestone that marks the beginning of an exciting journey in modern web development.

FastAPI is a modern, fast web framework for building APIs with Python based on standard Python type hints.

Key features that define FastAPI are the following:

- **Speed**: It's one of the fastest frameworks for building APIs in Python, thanks to its underlying Starlette framework for web parts and Pydantic for data handling

- **Ease of use**: FastAPI is designed to be easy to use, with intuitive coding that accelerates your development time

- **Automatic documentation**: With FastAPI, the API documentation is generated automatically, a feature that is both a time-saver and a boon for developers

How to do it...

We will now explore how to use those features effectively with some general guidance.

We will go through the following steps:

- Applying asynchronous programming to our existing endpoints to improve time efficiency

- Exploring routers and endpoints to better organize large code bases

- Running your first FastAPI server with a basic configuration

- Exploring the automatic documentation

Applying asynchronous programming

One of the most powerful features of FastAPI is its support for asynchronous programming. This allows your applications to handle more requests simultaneously, making them more efficient. Asynchronous programming is a style of concurrent programming in which tasks are executed without blocking the execution of other tasks, improving the overall performance of your application. To integrate asynchronous programming smoothly, FastAPI leverages the `async/await` syntax (`https://fastapi.tiangolo.com/async/`) and automatically integrates asynchronous functions.

So, the `read_root()` function in `main.py` from the previous code snippet in the *Creating a new FastAPI project* recipe can be written as follows:

```
@app.get("/")
async def read_root():
    return {"Hello": "World"}
```

In this case, the behavior of the code will be exactly the same as before.

Exploring routers and endpoints

In FastAPI, organizing your code into routers and endpoints is a fundamental practice. This organization helps in making your code cleaner and more modular.

Endpoints

Endpoints are the points at which API interactions happen. In FastAPI, an endpoint is created by decorating a function with an HTTP method, such as `@app.get("/")`.

This signifies a GET request to the root of your application.

Consider the following code snippet:

```python
from fastapi import FastAPI
app = FastAPI()


@app.get("/")
async def read_root():
    return {"Hello": "World"}
```

In this snippet, we define an endpoint for the root URL ("/"). When a GET request is made to this URL, the `read_root` function is invoked, returning a JSON response.

Routers

When we need to handle multiple endpoints that are in different files, we can benefit from using routers. Routers assist us in grouping our endpoints into different modules, which makes our code base easier to maintain and understand. For example, we could use one router for operations related to users and another for operations related to products.

To define a router, first create a new file in the `fastapi_start` folder called `router_example.py`. Then, create the router as follows:

```python
from fastapi import APIRouter

router = APIRouter()

@router.get("/items/{item_id}")
async def read_item(item_id: int):
    return {"item_id": item_id}
```

You can now reuse it and attach the router to the FastAPI server instance in `main.py`:

```python
import router_example
from fastapi import FastAPI

app = FastAPI()
```

```
app.include_router(router_example.router)


@app.get("/")
async def read_root():
    return {"Hello": "World"}
```

You now have the code to run the server that includes the router for the GET /items endpoint importer from another module.

Running your first FastAPI server

To run your FastAPI application, you need to point Uvicorn to your app instance. If your file is named main.py and your FastAPI instance is called app, you can start your server like this at the fastapi_start folder level:

```
$ uvicorn main:app --reload
```

The --reload flag makes the server restart after code changes, making it ideal for development.

Once the server is running, you can access your API at http://127.0.0.1:8000. If you visit this URL in your browser, you'll see the JSON response from the "/" endpoint we have just created.

Exploring the automatic documentation

One of the most exciting features of FastAPI is its automatic documentation. When you run your FastAPI application, two documentation interfaces are automatically generated: **Swagger UI** and **Redoc**.

You can access these at http://127.0.0.1:8000/docs for Swagger UI and http://127.0.0.1:8000/redoc for Redoc.

These interfaces provide an interactive way to explore your API and test its functionality.

See also

You can discover more about what we covered in the recipe at the following links:

- *First Steps*: https://fastapi.tiangolo.com/tutorial/first-steps/
- *Docs URLs*: https://fastapi.tiangolo.com/tutorial/metadata/#docs-urls
- *Concurrency and async / await*: https://fastapi.tiangolo.com/async/

Defining your first API endpoint

Now that you have a fundamental grasp of FastAPI and your development environment is all set up, it's time to take the next thrilling step: creating your first API endpoint.

This is where the real magic of FastAPI begins to shine. You'll see how effortlessly you can build a functional API endpoint, ready to respond to HTTP requests.

In this recipe, you will create a basic draft of a backend service for a bookstore.

Getting ready

Make sure you know how to start a basic FastAPI project from the *Creating a new FastAPI project* recipe.

How to do it...

In the realm of web APIs, the GET request is perhaps the most common. It's used to retrieve data from the server. In FastAPI, handling a GET request is simple and intuitive. Let's create a basic GET endpoint.

Imagine you're building an API for a bookstore. Your first endpoint will provide information about a book when given its ID. Here's how you do it:

1. Create a new `bookstore` folder that will contain the code you are going to write.

2. Create in it a `main.py` file containing the server instance:

```python
from fastapi import FastAPI


app = FastAPI()


@app.get("/books/{book_id}")
async def read_book(book_id: int):
    return {
        "book_id": book_id,
        "title": "The Great Gatsby",
        "author": "F. Scott Fitzgerald"
    }
```

In the preceding code snippet, the `@app.get("/books/{book_id}")` decorator tells FastAPI that this function will respond to GET requests at the `/books/{book_id}` path. `{book_id}` in the path is a path parameter, which you can use to pass values dynamically. FastAPI automatically extracts the `book_id` parameter and passes it to your function.

> **Type hints and automatic data validation**
>
> Notice the use of type hints (`book_id: int`). FastAPI uses these hints to perform data validation. If a request is made with a non-integer `book_id` parameter, FastAPI automatically sends a helpful error response.

How it works...

With your `GET` endpoint defined, run your FastAPI application using Uvicorn, just as you did previously:

```
$ uvicorn main:app --reload
```

On the terminal, you can read the message logs describing that the server is running on port `8000`.

One of FastAPI's most beloved features is its automatic generation of interactive API documentation using Swagger UI. This tool allows you to test your API endpoints directly from your browser without writing any additional code, and you can directly check the presence of the newly created endpoint in it.

Using Swagger UI

To test your new GET endpoint, navigate to `http://127.0.0.1:8000/docs` in your browser. This URL brings up the Swagger UI documentation for your FastAPI application. Here, you'll see your `/books/{book_id}` endpoint listed. Click on it, and you'll be able to execute a test request right from the interface. Try inputting a book ID and see the response your API generates.

Postman – a versatile alternative

While Swagger UI is convenient for quick tests, you might want to use a more robust tool such as Postman for more complex scenarios. Postman is an API client that lets you build, test, and document your APIs more extensively.

To use Postman, download and install it from Postman's website (`https://www.postman.com/downloads/`).

Once installed, create a new request. Set the method to `GET` and the request URL to your FastAPI endpoint, `http://127.0.0.1:8000/books/1`. Hit **Send**, and Postman will display the response from your FastAPI server.

Working with path and query parameters

One of the most crucial aspects of API development is handling parameters. Parameters allow your API to accept input from users, making your endpoints dynamic and responsive.

In this recipe, we will explore how to capture and handle path, query parameters, and test them efficiently, enhancing the flexibility and functionality of your FastAPI applications.

Getting ready

To follow the recipe, make sure you know how to create a basic endpoint from the previous recipe.

How to do it...

Path parameters are parts of the URL that are expected to change. For instance, in an endpoint such as /books/{book_id}, book_id is a path parameter. FastAPI allows you to capture these parameters effortlessly and use them in your function.

1. Let's expand our bookstore API with a new endpoint that uses path parameters. This time, we'll create a route to get information about a specific author:

```
@app.get("/authors/{author_id}")
async def read_author(author_id: int):
    return {
        "author_id": author_id,
        "name": "Ernest Hemingway"
    }
```

The name will not change; however, the author_id value will be the one provided by the query request.

Query parameters are used to refine or customize the response of an API endpoint. They can be included in the URL after a question mark (?). For instance, /books?genre=fiction&year=2010 might return only books that fall under the fiction genre released in 2010.

2. Let's add query parameters to our existing endpoint. Suppose we want to allow users to filter books by their publication year:

```
@app.get("/books")
async def read_books(year: int = None):
    if year:
        return {
            "year": year,
            "books": ["Book 1", "Book 2"]
        }
    return {"books": ["All Books"]}
```

Here, year is an optional query parameter. By assigning None as a default value, we make it optional. If a year is specified, the endpoint returns books from that year; otherwise, it returns all books.

> **Exercise**
> Using the `APIRouter` class, refactor each endpoint in a separate file and add the route to the FastAPI server.

How it works...

Now, from the command terminal, spin up the server with Uvicorn by running the following command:

```
$ uvicorn main:app
```

Testing endpoints with path parameters can be done using Swagger UI or Postman, similar to how we tested our basic `GET` endpoint.

In Swagger UI, at `http://localhost:8000/docs`, navigate to your `/authors/{author_id}` endpoint. You'll notice that it prompts you to enter an `author_id` value before you can try it out. Enter a valid integer and execute the request. You should see a response with the author's information.

The `GET /books` endpoint will now show an optional field for the `year` query parameter. You can test it by entering different years and observing the varying responses.

If you use Postman instead, create a new `GET` request with the `http://127.0.0.1:8000/authors/1` URL. Sending this request should yield a similar response.

In Postman, append the query parameter to the URL like so: `http://127.0.0.1:8000/books?year=2021`. Sending this request should return books published in the year 2021.

See also

You can find more about path and query parameters in the FastAPI official documentation at the following links:

- *Path Parameters*: `https://fastapi.tiangolo.com/tutorial/path-params/`
- *Query Parameters*: `https://fastapi.tiangolo.com/tutorial/query-params/`

Defining and using request and response models

In the world of API development, data handling is a critical aspect that determines the robustness and reliability of your application. FastAPI simplifies this process through its seamless integration with **Pydantic**, a data validation and settings management library using Python type annotations. The recipe will show you how to define and use request and response models in FastAPI, ensuring your data is well structured, validated, and clearly defined.

Pydantic models are a powerful feature for data validation and conversion. They allow you to define the structure, type, and constraints of the data your application handles, both for incoming requests and outgoing responses.

In this recipe, we will see how to use Pydantic to ensure that your data conforms to the specified schema, providing an automatic layer of safety and clarity.

Getting ready

This recipe requires you to know how to set up a basic endpoint in FastAPI.

How to do it...

We will break the process into the following steps:

1. Creating the model
2. Defining the request body
3. Validating request data
4. Managing response formats

Creating the model

Let's create a Pydantic `BaseModel` class for our bookstore application in a new file called `models.py`.

Suppose we want to have a model for a book that includes the title, author, and publication year:

```
from pydantic import BaseModel


class Book(BaseModel):
    title: str
    author: str
    year: int
```

Here, `Book` is a Pydantic `BaseModel` class with three fields: `title`, `author`, and `year`. Each field is typed, ensuring that any data conforming to this model will have these attributes with the specified data types.

Defining the request body

In FastAPI, Pydantic models are not just for validation. They also serve as the request body. Let's add an endpoint to our application where users can add new books:

```
from models import Book


@app.post("/book")
```

```
async def create_book(book: Book):
    return book
```

In this endpoint, when a user sends a POST request to the /book endpoint with JSON data, FastAPI automatically parses and validates it against the Book model. If the data is invalid, the user gets an automatic error response.

Validating request data

Pydantic offers advanced validation features. For instance, you can add regex validations, default values, and more:

```
from pydantic import BaseModel, Field

class Book(BaseModel):
    title: str = Field(..., min_length=1, max_length=100)
    author: str = Field(..., min_length=1, max_length=50)
    year: int = Field(..., gt=1900, lt=2100)
```

For an exhaustive list of validation features, have a look at Pydantic's official documentation: https://docs.pydantic.dev/latest/concepts/fields/.

Next, you can proceed to manage the response format.

Managing response formats

FastAPI allows you to define response models explicitly, ensuring that the data returned by your API matches a specific schema. This can be particularly useful for filtering out sensitive data or restructuring the response.

For example, let's say you want the /allbooks GET endpoint to return a list of books, but only with their titles and authors, omitting the publication year. In main.py, add the following accordingly:

```
from pydantic import BaseModel

class BookResponse(BaseModel):
    title: str
    author: str

@app.get("/allbooks")
async def read_all_books()    ➤ list[BookResponse]:
    return [
        {
            "id": 1,
```

```
            "title": "1984",
            "author": "George Orwell"},
        {
            "id": 1,
            "title": "The Great Gatsby",
            "author": "F. Scott Fitzgerald",
        },
    ]
```

Here, the `-> list[BookResponse]` function type hint tells FastAPI to use the BookResponse model for responses, ensuring that only the title and author fields are included in the response JSON. Alternatively, you can specify the response type in the endpoint decorator's arguments as follows:

```
@app.get("/allbooks", response_model= list[BookResponse])
async def read_all_books() -> Any:
# rest of the endpoint content
```

The `response_model` argument takes priority and can be used instead of the type hint to resolve type checker issues that may occur.

Check the documentation at `http://127.0.0.1:8000/docs`. Unroll the `/allbooks` endpoint details, and you will notice the example value response based on the schema as follows:

```
[
    {
      "title": "string",
      "author": "string"
    }
]
```

By mastering Pydantic models in FastAPI, you are now capable of handling complex data structures with ease and precision. You've learned to define request bodies and manage response formats, ensuring data consistency and integrity throughout your application.

See also

Pydantic is a standalone project largely used for data validation in Python with many more features than what the recipe has shown. Feel free to have a look at the official documentation at the following link:

- *Pydantic*: `https://docs.pydantic.dev/latest/`

You can see more on response model usage at the FastAPI official documentation link:

- *Response Model - Return Type*: `https://fastapi.tiangolo.com/tutorial/
  response-model/`

Handling errors and exceptions

Error handling is an essential aspect of developing robust and reliable web applications. In FastAPI, managing errors and exceptions is not just about catching unexpected issues but also about proactively designing your application to respond to various error scenarios gracefully.

This recipe will guide you through custom error handling, validating data and handling exceptions, and testing these scenarios to ensure your FastAPI applications are resilient and user-friendly.

How to do it...

FastAPI provides built-in support for handling exceptions and errors.

When an error occurs, FastAPI returns a JSON response containing details about the error, which is very useful for debugging. However, there are situations where you might want to customize these error responses for better user experience or security.

Let's create a custom error handler that catches a specific type of error and returns a custom response. For instance, if a requested resource is not found, you might want to return a more friendly error message.

To do it, in the `main.py` file, add the following code accordingly:

```
from fastapi import FastAPI, HTTPException
from starlette.responses import JSONResponse

@app.exception_handler(HTTPException)
async def http_exception_handler(request, exc):
    return JSONResponse(
        status_code=exc.status_code,
        content={
            "message": "Oops! Something went wrong"
        },
    )
```

In this example, the `http_exception_handler` function will be used to handle `HTTPException` errors. Whenever an `HTTPException` error is raised anywhere in your application, FastAPI will use this handler to return a custom response.

You can test the response by creating a new endpoint that raises an HTTP exception:

```python
@app.get("/error_endpoint")
async def raise_exception():
    raise HTTPException(status_code=400)
```

The endpoint will explicitly throw the HTTP error response to showcase the customized message defined in the previous step.

Now, spin the server from the command line with the following command:

```
$ uvicorn main:app
```

Open the browser at http://localhost:8000/error_endpoint, and you will have a JSON response like this:

```json
{
    "message": "Oops! Something went wrong"
}
```

The response returns the default message we defined for any HTTP exception returned by the code.

There's more...

As discussed in the previous recipe, *Defining and using request and response models*, FastAPI uses Pydantic models for data validation. When a request is made with data that does not conform to the defined model, FastAPI automatically raises an exception and returns an error response.

In some cases, you might want to customize the response for validation errors. FastAPI makes this quite straightforward:

```python
import json

from fastapi import Request, status
from fastapi.exceptions import RequestValidationError
from fastapi.responses import PlainTextResponse

@app.exception_handler(RequestValidationError)
async def validation_exception_handler(
    request: Request,
    exc: RequestValidationError
):
    return PlainTextResponse(
        "This is a plain text response:"
        f" \n{json.dumps(exc.errors(), indent=2)}",
```

```
        status_code=status.HTTP_400_BAD_REQUEST,
    )
```

This custom handler will catch any `RequestValidationError` error and return a plain text response with the details of the error.

If you try, for example, to call the `POST /book` endpoint with a number type of `title` instead of a string, you will get a response with a status code of `400` and body:

```
This is a plain text response:
[
  {
    "type": "string_type",
    "loc": [
      "body",
      "author"
    ],
    "msg": "Input should be a valid string",
    "input": 3,
    "url": "https://errors.pydantic.dev/2.5/v/string_type"
  },
  {
    "type": "greater_than",
    "loc": [
      "body",
      "year"
    ],
    "msg": "Input should be greater than 1900",
    "input": 0,
    "ctx": {
      "gt": 1900
    },
    "url": "https://errors.pydantic.dev/2.5/v/greater_than"
  }
]
```

You can also, for example, mask the message to add a layer of security to protect from unwanted users using it incorrectly.

This is all you need to customize responses when a request validation error occurs.

You will use this basic knowledge as you move to the next chapter. *Chapter 2* will teach you more about data management in web applications, showing you how to set up and use SQL and NoSQL databases and stressing data security. This will not only improve your technical skills but also increase your awareness of creating scalable and reliable FastAPI applications.

See also

You can find more information about customizing errors and exceptions using FastAPI in the official documentation:

- *Handling Errors*: `https://fastapi.tiangolo.com/tutorial/handling-errors/`

2

Working with Data

Data handling is the backbone of any web application, and this chapter is dedicated to mastering this critical aspect. You will embark on a journey of working with data in FastAPI, where you'll learn the intricacies of integrating, managing, and optimizing data storage using both **Structured Query Language (SQL)** and **NoSQL** databases. We'll cover how FastAPI, combined with powerful database tools, can create efficient and scalable data management solutions.

Starting with SQL databases, you'll get hands-on experience in setting up a database, implementing **create, read, update and delete (CRUD)** operations, and understanding the nuances of working with SQLAlchemy – a popular **object-relational mapping (ORM)** option for Python. We'll then shift gears to NoSQL databases, delving into the world of **MongoDB.** You'll learn how to integrate it with FastAPI, handle dynamic data structures, and leverage the flexibility and scalability of NoSQL solutions.

But it's not just about storing and retrieving data. This chapter also focuses on best practices for securing sensitive data and managing transactions and concurrency in your databases. You'll explore how to protect your data from vulnerabilities and ensure the integrity and consistency of your application's data operations.

By the end of this chapter, you'll not only have a solid understanding of how to work with various database systems in FastAPI but also the skills to build robust and secure data models for your web applications. Whether it's implementing complex queries, optimizing database performance, or ensuring data security, this chapter provides the tools and knowledge you need to manage your application's data effectively.

In this chapter, we're going to cover the following recipes:

- Setting up SQL databases
- Understanding CRUD operations with SQLAlchemy
- Integrating MongoDB for NoSQL data storage
- Working with data validation and serialization
- Working with file uploads and downloads

- Handling asynchronous data operations
- Securing sensitive data and best practices

Each topic is designed to equip you with the necessary skills and knowledge to handle data in FastAPI efficiently, ensuring your applications are not only functional but also secure and scalable.

Technical requirements

To effectively run and understand the code in this chapter, ensure you have the following set up. If you've followed *Chapter 1, First Steps with FastAPI*, you should already have some of these installed:

- **Python**: Make sure you've installed Python version 3.9 or higher on your computer.

- **FastAPI**: Install FastAPI along with all its dependencies using the `pip install fastapi[all]` command. As we saw in *Chapter 1, First Steps with FastAPI*, this command also installs **Uvicorn**, an ASGI server that's necessary to run your FastAPI application.

- **Integrated development environment** (**IDE**): A suitable IDE such as **VS Code** or **PyCharm** should be installed. These IDEs offer excellent support for Python and FastAPI development, providing features such as syntax highlighting, code completion, and easy debugging.

- **MongoDB**: For the NoSQL database portions of this chapter, MongoDB needs to be installed on your local machine. Download and install the free community version server suitable for your operating system from `https://www.mongodb.com/try/download/community`.

  Make sure that MongoDB is correctly installed by running from the command line the Mongo Deamon:

  ```
  $ mongod --version
  ```

  You can check the MongoDB version installed on your machine by looking at the output after the installation. However, if you use Windows to run your MongoDB instance, the binary file daemon should be in a path location similar to `C:\Program>Files\MongoDB\Server\7.0\bin`. You need to open the terminal in this location to run the daemon or run:

  ```
  $ C:\Program>Files\MongoDB\Server\7.0\bin\mongod -- version
  ```

- **MongoDB tools**: While optional, tools such as the **MongoDB Shell** (`https://www.mongodb.com/try/download/shell`) and **MongoDB Compass GUI** (`https://www.mongodb.com/try/download/compass`) can greatly enhance your interaction with the MongoDB server. They provide a more user-friendly interface for managing databases, running queries, and visualizing data structures.

All the code and examples used throughout this chapter are available on GitHub for reference and download. Visit `https://github.com/PacktPublishing/FastAPI-Cookbook/tree/main/Chapter02` on GitHub to access the repository.

Setting up SQL databases

In the world of data handling, the power of Python meets the efficiency of SQL databases. This recipe aims to introduce you to how to integrate SQL databases within your application, a crucial skill for any developer looking to build robust and scalable web applications.

SQL is the standard language for managing and manipulating relational databases. When combined with FastAPI, it unlocks a world of possibilities in data storage and retrieval.

FastAPI's compatibility with SQL databases is facilitated through ORMs. The most popular one is **SQLAlchemy**. We will focus on it in this recipe.

Getting ready

To begin, you'll need to have FastAPI and SQLAlchemy installed in your virtual environment. If you followed the steps in *Chapter 1*, *First Steps with FastAPI*, you should have FastAPI already set up. For SQLAlchemy, a simple `pip` command is all that's needed:

```
$ pip install sqlalchemy
```

Once installed, the next step is to configure SQLAlchemy so that it can work with FastAPI. This involves setting up the database connection – a process we will walk through step by step.

How to do it...

Now, let's dive deeper into configuring SQLAlchemy for your FastAPI application. SQLAlchemy acts as the bridge between your Python code and the database, allowing you to interact with the database using Python classes and objects rather than writing raw SQL queries.

After installing SQLAlchemy, the next step is to configure it within your FastAPI application. This involves defining your database models – a representation of your database tables in Python code. In SQLAlchemy, models are typically defined using classes, with each class corresponding to a table in the database, and each attribute of the class corresponding to a column in the table.

Follow these steps to go through the process.

1. Create a new folder called `sql_example`, move inside it, and then create a file called `database.py` there. Write a `base` class to be used as a reference:

    ```
    from sqlalchemy.orm import DeclarativeBase


    class Base(DeclarativeBase):
        pass
    ```

To define a model in SQLAlchemy, you need to create a base class that derives from the `DeclarativeBase` class. This `Base` class maintains a catalog of classes and tables you've defined and is central to SQLAlchemy's ORM functionality.

You can learn more by reading the official documentation: `https://docs.sqlalchemy.org/en/13/orm/extensions/declarative/index.html`.

2. Once you have your `Base` class, you can start defining your models. For instance, if you have a table for users, your model might look something like this:

```
from sqlalchemy.orm import (
    Mapped,
    mapped_column
)


class User(Base):
    __tablename__ = "user"
    id: Mapped[int] = mapped_column(
        primary_key=True,
    )
    name: Mapped[str]
    email: Mapped[str]
```

In this model, `User` class corresponds to a table named `user` in the database, with columns for `id`, `name`, and `email`. Each `class attribute` specifies the data type of the column.

3. Once your models have been defined, the next step is to connect to the database and create these tables. SQLAlchemy uses a connection string to define the details of the database it needs to connect to. The format of this connection string varies depending on the database system you are using.

For example, a connection string for a SQLite database might look like this:

```
DATABASE_URL = "sqlite:///./test.db"
```

SQLite is a lightweight, file-based database that doesn't require a separate server process. It's an excellent choice for development and testing.

4. No further setup is required for SQLite as it will automatically create the `test.db` database file the first time you connect to it.

You will use the DATABASE_URL connection string to create an `Engine` object in SQLAlchemy that represents the core interface to the database:

```
from sqlalchemy import create_engine

engine = create_engine(DATABASE_URL)
```

5. With the engine created, you can proceed to create your tables in the database. You can do this by passing your base class and the engine to SQLAlchemy's create_all method:

```
Base.metadata.create_all(bind=engine)
```

Now that you've defined all the abstractions of the database in your code, you can proceed with setting the database connection.

Establishing a database connection

The final part of setting up a SQL database setup is establishing a database connection. This connection allows your application to communicate with the database, executing queries and retrieving data.

Database connections are managed with sessions. A session in SQLAlchemy represents a *workspace* for your objects, a place where you can add new records or fetch existing ones. Each session is bound to a single database connection.

To manage sessions, we need to create a SessionLocal class. This class will be used to create and manage session objects for the interactions with the database. Here's how you can create it:

```
from sqlalchemy.orm import sessionmaker

SessionLocal = sessionmaker(
    autocommit=False, autoflush=False, bind=engine
)
```

The sessionmaker function creates a factory for sessions. The autocommit and autoflush parameters are set to False, meaning you have to manually commit transactions and manage them when your changes are flushed to the database.

With the SessionLocal class in place, you can create a function that will be used in your FastAPI route functions to get a new database session. We can create it in the main.py module like so:

```
from database import SessionLocal

def get_db():
    db = SessionLocal()
    try:
        yield db
    finally:
        db.close()
```

In your route functions, you can use this function as a dependency to communicate with the database.

In FastAPI, this can be done with the Depends class. In the main.py file, you can then add an endpoint:

```
from fastapi import Depends, FastAPI
from sqlalchemy.orm import Session
from database import SessionLocal

app = FastAPI()

@app.get("/users/")
def read_users(db: Session = Depends(get_db)):
    users = db.query(User).all()
    return users
```

This approach ensures that a new session is created for each request and closed when the request is finished, which is crucial for maintaining the integrity of your database transactions.

You can then run the server with the following command:

```
$ uvicorn main:app --reload
```

If you try to call the endpoint GET at localhost:8000/users you will get an empty list since no users have been added already.

See also

You can discover more on how to set up a session in **SQLAlchemy** on the documentation page:

- *SQLAlchemy session*: https://docs.sqlalchemy.org/en/20/orm/session_basics.html

Understanding CRUD operations with SQLAlchemy

After setting up your SQL database with FastAPI, the next crucial step is creating database models. This process is central to how your application interacts with the database. **Database models** in SQLAlchemy are essentially Python classes that represent tables in your SQL database. They provide a high-level, object-oriented interface to manipulate database records as if they were regular Python objects.

In this recipe, we will set up the **create, read, update and delete** (CRUD) endpoints to interact with the database.

Getting ready

With the models set up, you can now implement CRUD operations. These operations form the backbone of most web applications, allowing you to interact with the database.

How to do it...

For each operation, we will create a dedicated endpoint implementing the interacting operation with the database.

Creating a new user

To add a new user, we'll use a POST request. In the main.py file, we must define an endpoint that receives user data, creates a new User instance in the body request, and adds it to the database:

```python
class UserBody(BaseModel):
    name: str
    email: str


@app.post("/user")
def add_new_user(
    user: UserBody,
    db: Session = Depends(get_db)
):
    new_user = User(
        name=user.name,
        email=user.email
    )
    db.add(new_user)
    db.commit()
    db.refresh(new_user)
    return new_user
```

In a few lines, you've created the endpoint to add a new user to the database.

Reading a specific user

To get a single user, we are going to use a GET endpoint:

```python
from fastapi import HTTPException


@app.get("/user")
def get_user(
```

```
    user_id: int,
    db: Session = Depends(get_db)
):
    user = (
        db.query(User).filter(
            User.id == user_id
        ).first()
    )
    if user is None:
        raise HTTPException(
            status_code=404,
            detail="User not found"
        )

    return user
```

The endpoint will return a 404 response status if the user does not exist.

Updating a user

Updating a record via an API offers various approaches, including PUT, PATCH, or POST methods. Despite theoretical nuances, the choice of method often boils down to personal preference. I favor using a POST request and augmenting the /user endpoint with a user_id parameter. This simplifies the process, minimizing the need for extensive memorization. You can integrate this endpoint within the main.py module like so:

```
@app.post("/user/{user_id}")
def update_user(
    user_id: int,
    user: UserBody,
    db: Session = Depends(get_db),
):
    db_user = (
        db.query(User).filter(
            User.id == user_id
        ).first()
    )
    if db_user is None:
        raise HTTPException(
            status_code=404,
            detail="User not found"
        )
```

```
    db_user.name = user.name
    db_user.email = user.email
    db.commit()
    db.refresh(db_user)
    return db_user
```

This is all you need to do to create the endpoint to update a user record in the database.

Deleting a user

To conclude, deleting a user within the same `main.py` module involves utilizing a `DELETE` request, as shown here:

```
@app.delete("/user")
def delete_user(
    user_id: int, db: Session = Depends(get_db)
):
    db_user = (
        db.query(User).filter(
            User.id == user_id
        ).first()
    )
    if db_user is None:
        raise HTTPException(
            status_code=404,
            detail="User not found"
        )
    db.delete(db_user)
    db.commit()
    return {"detail": "User deleted"}
```

These endpoints cover the basic CRUD operations and demonstrate how FastAPI can be integrated with SQLAlchemy for database operations. By defining these endpoints, your application can create, retrieve, update, and delete user data, providing a fully functional API for client interactions.

Now that you have implemented all the operations, you can spin the server by running:

```
$ uvicorn main:app
```

Then open the inreactive documentation at `http://localhost:8000/docs` and start playing with the endpoints by creating, reading, updating and deleting users.

Mastering these CRUD operations in FastAPI is a significant step in building dynamic and data-driven web applications. With the knowledge of how to integrate FastAPI endpoints with SQLAlchemy models, you are well-equipped to develop complex and efficient applications.

See also

You can find a clear quick start on how to set up the ORM for CRUD operations with SQLAlchemy on the official documentation page:

- *SQLAlchemy ORM Quick Start*: `https://docs.sqlalchemy.org/en/20/orm/quickstart.html`

Integrating MongoDB for NoSQL data storage

Transitioning from SQL to NoSQL databases opens up a different paradigm in data storage and management. **NoSQL databases**, like MongoDB, are known for their flexibility, scalability, and ability to handle large volumes of unstructured data. In this recipe, we'll explore how to integrate MongoDB, a popular NoSQL database, with FastAPI.

NoSQL databases differ from traditional SQL databases in that they often allow for more dynamic and flexible data models. MongoDB, for example, stores data in **binary JSON (BSON)** format, which can easily accommodate changes in data structure. This is particularly useful in applications that require rapid development and frequent updates to the database schema.

Getting ready

Make sure you've installed MongoDB on your machine. If you haven't done it yet, you can download the installer from `https://www.mongodb.com/try/download/community`.

FastAPI doesn't provide a built-in ORM for NoSQL databases. However, integrating MongoDB into FastAPI is straightforward thanks to Python's powerful libraries.

We'll use `pymongo`, a Python package driver to interact with MongoDB.

First, ensure you have MongoDB installed and running on your machine.

Then, you can install `pymongo` with `pip`:

```
$ pip install pymongo
```

With `pymongo` installed, we can now establish a connection to a MongoDB instance and start performing database operations.

How to do it...

We can quickly connect our application to a Mongo DB instance running on our local machine by by applying the following steps.

1. Create a new project folder called nosql_example. Start by defining connection configuration in a database.py file:

    ```
    From pymongo import MongoClient

    client = MongoClient()

    database = client.mydatabase
    ```

    In this example, mydatabase is the name of your database. You can replace it with the name you prefer. Here, MongoClient establishes a connection to a MongoDB instance running locally on the *default port of 27017.*

2. Once the connection has been set up, you can define your collections (equivalent to tables in SQL databases) and start interacting with them. MongoDB stores data in collections of documents, where each document is a JSON-like structure:

    ```
    user_collection = database["users"]
    ```

    Here, user_collection is a reference to the users collection in your MongoDB database.

3. To test the connection, you can create an endpoint that will retrieve all users that should return an empty list in a main.py file:

    ```
    from database import user_collection
    from fastapi import FastAPI, HTTPException
    from pydantic import BaseModel

    app = FastAPI()


    class User(BaseModel):
        name: str
        email: str

    @app.get("/users")
    def read_users() -> list[User]:
        return [user for user in user_collection.find()]
    ```

4.   Now, run your mongod instance. You can do it from the command line:

```
$ mongod
```

If you run on Windows the command will be:

```
$ C:\Program>Files\MongoDB\Server\7.0\bin\mongod
```

And that's it. To test it, in a separate terminal window, spin up the FastAPI server by running the following command:

```
$ uvicorn main:app
```

Then, simply open your browser at http://localhost:8000/users; you will get an empty list. This means that your database connection is correctly working.

Now that the connection has been set up, we are going to create an endpoint to add a user and one to retrieve a specific user with an ID. We'll create both endpoints in the main.py module.

Creating a new user

To add a new document to a collection, use the insert_one method:

```python
class UserResponse(User):
    id: str


@app.post("/user")
def create_user(user: User):
    result = user_collection.insert_one(
        user.model_dump(exclude_none=True)
    )
    user_response = UserResponse(
        id=str(result.inserted_id),
        *user.model_dump()
    )
    return user_response
```

The endpoint we've just created returns the affected id number in the response to be used as input for other endpoints.

Reading a user

To retrieve a document, you can use the find_one method:

```python
from bson import ObjectId

@app.get("/user")
```

```python
def get_user(user_id: str):
    db_user = user_collection.find_one(
        {
            "_id": ObjectId(user_id)
            if ObjectId.is_valid(user_id)
            else None
        }
    )
    if db_user is None:
        raise HTTPException(
            status_code=404,
            detail="User not found"
        )
    user_response = UserResponse(
        id=str(db_user["_id"]), **db_user
    )
    return user_response
```

If the user with the specified ID doesn't exist, it will return a response status of 404.

In Mongo, the ID of the document is not stored in plain text, but in a 12-byte object. That's why we need to initialize a dedicated `bson.ObjectId` when querying the database and explicitly decode to `str` when returning the value through the response.

You can then spin up the server with `uvicorn`:

```
$ uvicorn main:app
```

You can see the endpoints on the interactive documentation page: `http://localhost:8000/docs`. Ensure you test every endpoint and the interaction among them thoroughly.

By integrating MongoDB with FastAPI, you gain the ability to handle dynamic, schemaless data structures, which is a significant advantage in many modern web applications. This recipe has equipped you with the knowledge to set up MongoDB, define models and collections, and perform CRUD operations, providing a solid foundation for building versatile and scalable applications with FastAPI and MongoDB.

See also

You can dig into how to use the **PyMongo** Python client by reading the official documentation:

- *PyMongo documentation*: `https://pymongo.readthedocs.io/en/stable/`

Working with data validation and serialization

Effective data validation stands as a cornerstone of robust web applications, ensuring that incoming data meets predefined criteria and remains safe for processing.

FastAPI harnesses the power of Pydantic, a Python library dedicated to data validation and serialization. By integrating Pydantic models, FastAPI streamlines the process of validating and serializing data, offering an elegant and efficient solution. This recipe shows how to utilize Pydantic models within FastAPI applications, exploring how they enable precise validation and seamless data serialization.

Getting ready

Pydantic models are essentially Python classes that define the structure and validation rules of your data. They use Python's type annotations to validate that incoming data matches the expected format. When you use a Pydantic model in your FastAPI endpoints, FastAPI automatically validates incoming request data against the model.

In this recipe, we're going to use Pydantic's email validator, which comes with the default `pydantic` package distribution. However, it needs to be installed in your environment. You can do this by running the following command:

```
$ pip install pydantic[email]
```

Once the package has been installed, you are ready to start this recipe.

How to do it...

Let's use it in the previous project. In the `main.py` module, we'll modify the `UserCreate` class, which is used to accept only valid `email` fields:

```
from typing import Optional
from pydantic import BaseModel, EmailStr

class UserCreate(BaseModel):
    name: str
    email: EmailStr
```

In this model, `name` is a required string and `email` must be a valid email address. FastAPI will automatically use this model to validate incoming data for any endpoint that expects a `UserCreate` object.

Let's say you try to add a user at the POST /user endpoint with an invalid user information body, as shown here:

```
{
    "name": "John Doe",
    "email": "invalidemail.com",
}
```

You will get a 422 response with a message body specifying the invalid fields.

Serialization and deserialization concepts

Serialization is the process of converting complex data types, such as Pydantic models or database models, into simpler formats such as JSON, which can be easily transmitted over the network. **Deserialization** is the reverse process, converting incoming data into complex Python types.

FastAPI handles serialization and deserialization automatically using Pydantic models. When you return a Pydantic model from an endpoint, FastAPI serializes it to JSON. Conversely, when you accept a Pydantic model as an endpoint parameter, FastAPI deserializes the incoming JSON data into the model.

For example, the get_user endpoint from the NoSQL example can be improved further like so:

```
class UserResponse(User):
    id: str


@app.get("/user")
def get_user(user_id: str) -> UserResponse:
    db_user = user_collection.find_one(
        {
            "_id": ObjectId(user_id)
            if ObjectId.is_valid(user_id)
            else None
        }
    )
    if db_user is None:
        raise HTTPException(
            status_code=404,
            detail="User not found"
        )
    db_user["id"] = str(db_user["_id"])
    return db_user
```

In this endpoint, FastAPI deserializes the incoming JSON data into a User object and then serializes the returned UserResponse object back into JSON.

This automatic serialization and deserialization make working with JSON data in FastAPI straightforward and type-safe.

Advanced validation techniques

Pydantic offers a range of advanced validation techniques that you can leverage in FastAPI. These include custom validators and complex data types.

Custom validators allow you to define complex validation logic for your models. You can create a custom validator by adding a method to your Pydantic model decorated with `@field_validator`.

For example, you could add a validator to ensure that a user's age is within a certain range:

```python
from pydantic import BaseModel, EmailStr, field_validator


class User(BaseModel):
    name: str
    email: EmailStr
    age: int

    @field_validator("age")
    def validate_age(cls, value):
        if value < 18 or value > 100:
            raise ValueError(
                "Age must be between 18 and 100"
            )
        return value
```

This validator ensures that the `age` field of the `User` model is between `18` and `100`.

If the validation fails, a descriptive error message is automatically returned to the client.

Pydantic also supports **complex data types** such as `list`, `dict`, and custom types, allowing you to define models that closely represent your data structures.

For instance, you can have a model with a list of items:

```python
class Tweet(BaseModel):
    content: str
    hashtags: list[str]


class User(BaseModel):
    name: str
```

```
        email: EmailStr
        age: Optional[int]
        tweets: list[Tweet] | None = None
```

In this example, the User model has an optional tweets field, which is a list of Tweet objects.

By leveraging Pydantic's advanced validation features, you can ensure that the data your FastAPI application processes is not only in the correct format but also adheres to your specific business logic and constraints. This provides a robust and flexible way to handle data validation and serialization in your FastAPI applications.

See also

You can learn more about the potential of Pydantic validators on the documentation page:

- *Pydantic validators*: https://docs.pydantic.dev/latest/concepts/validators/

Working with file uploads and downloads

Handling files is a common requirement in web applications, whether it's uploading user avatars, downloading reports, or processing data files. FastAPI provides efficient and easy-to-implement methods for both uploading and downloading files. This recipe will guide you through how to set up and implement file handling in FastAPI.

Getting ready

Let's create a new project directory called uploads_and_downloads that contains a main. py module with a folder called uploads. This will contain the files from the application side. The directory structure will look like this:

```
uploads_and_downloads/
|— uploads/
|— main.py
```

We can now proceed to create the appropriate endpoints.

How to do it...

To handle file uploads in FastAPI, you must use the File and UploadFile classes from FastAPI. The UploadFile class is particularly useful as it provides an asynchronous interface and spools large files to disk to avoid memory exhaustion.

In the `main.py` module, you can define the endpoint to upload files like so:

```
from fastapi import FastAPI, File, UploadFile

app = FastAPI()

@app.post("/uploadfile")
async def upload_file(
    file: UploadFile = File(...)):
    return {"filename": file.filename}
```

In this example, `upload_file` is an endpoint that accepts an uploaded file and returns its filename. The file is received in the form of an `UploadFile` object, which you can then save to disk or process further.

Implementing file uploads

When implementing file uploads, it's essential to handle the file data correctly to ensure it is saved without corruption. Here's an example of how you can save the uploaded file to a directory on your server.

Create a new folder project called `uploads_downloads`.

In the `main.py` module, create the `upload_file` endpoint:

```
import shutil

from fastapi import FastAPI, File, UploadFile

app = FastAPI()


@app.post("/uploadfile")
async def upload_file(
    file: UploadFile = File(...),
):
    with open(
        f"uploads/{file.filename}", "wb"
    ) as buffer:
        shutil.copyfileobj(file.file, buffer)

    return {"filename": file.filename}
```

This code snippet opens a new file in write-binary mode in the `uploads` directory and uses `shutil.copyfileobj` to copy the file content from the `UploadFile` object to the new file.

> **Important note**
>
> In a production environment, remember to handle exceptions and errors appropriately, especially for larger files

Create a text file called `content.txt` with some text in it.

Start the server by running the `uvicorn main:app` command. Then, access the interactive documentation; you'll observe that the endpoint we just created for file uploads includes a mandatory field prompting users to upload a file. Upon testing the endpoint by uploading a file, you'll discover the uploaded file residing within the designated `uploads` folder.

Managing file downloads and storage

Downloading files is the reverse process of uploading. In FastAPI, you can easily set up an endpoint to serve files for download. The `FileResponse` class is particularly useful for this purpose. It streams files from the server to the client, making it efficient for serving large files.

Here's a simple file download endpoint:

```
from fastapi.responses import FileResponse


@app.get(
    "/downloadfile/{filename}",
    response_class=FileResponse,
)
async def download_file(filename: str):
    if not Path(f"uploads/{filename}").exists():
        raise HTTPException(
            status_code=404,
            detail=f"file {filename} not found",
        )

    return FileResponse(
        path=f"uploads/{filename}", filename=filename
    )
```

In this example, `download_file` is an endpoint that serves files from the `uploads` directory for download. Here, `FileResponse` automatically sets the appropriate content-type header based on the file type and handles streaming the file to the client.

The content of the file will be the response body of the endpoint.

Handling file storage is another crucial aspect, especially when dealing with a large number of files or large file sizes. It's often advisable to store files in a dedicated file storage system rather than directly on your web server. Cloud storage solutions such as **Amazon S3**, **Google Cloud Storage**, or **Azure Blob Storage** can be integrated into your FastAPI application for scalable and secure file storage. Additionally, consider implementing cleanup routines or archival strategies to manage the life cycle of the files you store.

See also

You can learn more about how to manage uploaded files on the official documentation page:

- *FastAPI request files*: `https://fastapi.tiangolo.com/tutorial/request-files/`

Handling asynchronous data operations

Asynchronous programming is a core feature of FastAPI that allows you to develop highly efficient web applications. It allows your application to handle multiple tasks concurrently, making it particularly well-suited for I/O-bound operations, such as database interactions, file handling, and network communication.

Let's delve into leveraging asynchronous programming in FastAPI for data operations, enhancing the performance and responsiveness of your applications.

Getting ready

FastAPI is built on Starlette and Pydantic, which provide a robust foundation for writing asynchronous code in Python using the `asyncio` library with `async/await` syntax.

The `asyncio` library allows you to write non-blocking code that can pause its execution while waiting for I/O operations to complete, and then resume where it left off, all without blocking the main execution thread.

This recipe demonstrates the benefits of using `asyncio` with FastAPI in a simple, practical example.

How to do it...

Let's create an application with two endpoints, one that runs a sleeping operation, the other that run the sleeping operation as well but in asynchrounous mode. Create a new project folder called `async_example` containing the `main.py` module. Fill the module as follows.

1. Let's start by creating the FastAPI server object class:

```
from fastapi import FastAPI

app = FastAPI()
```

2. Now, let's create an endpoint that sleeps for 1 second:

```
import time

@app.get("/sync")
def read_sync():
    time.sleep(2)
    return {
        "message": "Synchrounouns blocking endpoint"
    }
```

The sleeping operation represents the waiting time to get a response from the database in a real-life scenario.

3. Now, let's create the same endpoint for the `async def` version. The sleeping operation will be the sleep function from the `asyncio` module:

```
import asyncio

@app.get("/async")
async def read_async():
    await asyncio.sleep(2)
    return {
        "message":
        "Asynchronous non-blocking endpoint"
    }
```

Now, we have two endpoints, GET /sync and GET/async, that are similar except for the fact the second contains a non-blocking sleeping operation.

Once we have our application with the endpoints, let's create a separate Python script to measure the time to serve a traffic demand. Let's call it `timing_api_calls.py` and start building it through the following steps.

1. Let's define the function to run the server:

```
import uvicorn
from main import app

def run_server():
    uvicorn.run(app, port=8000, log_level="error")
```

2. Now, let's define the start of the server as a context manager:

```
from contextlib import contextmanager
from multiprocessing import Process

@contextmanager
def run_server_in_process():
    p = Process(target=run_server)
    p.start()
    time.sleep(2)   # Give the server a second to start
    print("Server is running in a separate process")
    yield
    p.terminate()
```

3. Now, we can define a function that makes *n* concurrent requests to a specified path endpoint:

```
async def make_requests_to_the_endpoint(
    n: int, path: str
):
    async with AsyncClient(
        base_url="http://localhost:8000"
    ) as client:
        tasks = (
            client.get(path, timeout=float("inf"))
            for _ in range(n)
        )

        await asyncio.gather(*tasks)
```

4. At this point, we can gather the operations into a main function, make *n* calls for each of the endpoints, and print the time to serve all the calls to the terminal:

```
async def main(n: int = 10):
    with run_server_in_process():

        begin = time.time()
        await make_requests_to_the_endpoint(n,
                                        "/sync")
        end = time.time()
        print(
            f"Time taken to make {n} requests "
            f"to sync endpoint: {end - begin} seconds"
```

```
    )

    begin = time.time()
    await make_requests_to_the_endpoint(n,
                                        "/async")
    end = time.time()
    print(
        f"Time taken to make {n} requests "
        f"to async endpoint: {end - begin}
        seconds"
    )
```

5. Finally, we can run the main function in the `asyncio` event loop:

```
if __name__ == "__main__":
    asyncio.run(main())
```

Now that we have built our timing script, let's run it from the command terminal as follows:

```
$ python timing_api_calls.py
```

If you keep the default number of calls set to 10, your output will likely resemble the one on my machine:

```
Time taken to make 10 requests to sync endpoint: 2.3172452449798584
seconds
Time taken to make 10 requests to async endpoint: 2.3033862113952637
seconds
```

It looks like there is no improvement at all with using asyncio programming.

Now, try to set the number of calls to 100:

```
if __name__ == "__main__":
    asyncio.run(main(n=100))
```

The output will likely be more like this:

```
Time taken to make 100 requests to sync endpoint: 6.424988269805908
seconds
Time taken to make 100 requests to async endpoint: 2.423431873321533
seconds
```

This improvement is certainly noteworthy, and it's all thanks to the use of asynchronous functions.

There's more...

Asynchronous data operations can significantly improve the performance of your application, particularly when dealing with high-latency operations such as database access. By not blocking the main thread while waiting for these operations to complete, your application remains responsive and capable of handling other incoming requests or tasks.

If you already wrote CRUD operations synchronously, as we did in the previous recipe, *Understanding CRUD operations with SQLAlchemy*, implementing asynchronous CRUD operations in FastAPI involves modifying your standard CRUD functions so that they're asynchronous with the `sqlalchemy [asyncio]` library. Similarly to SQL, for NoSQL, you will need to use the `motor` package, which is the asynchronous MongoDB client built on top of `pymongo`.

However, it's crucial to use asynchronous programming judiciously. Not all parts of your application will benefit from asynchrony, and in some cases, it can introduce complexity. Here are some best practices for using asynchronous programming in FastAPI:

- **Use Async for I/O-bound operations**: Asynchronous programming is most beneficial for I/O-bound operations (such as database access, file operations, and network requests). CPU-bound tasks that require heavy computation might not benefit as much from asynchrony.

- **Database transactions**: When working with databases asynchronously, be mindful of transactions. Ensure that your transactions are correctly managed to maintain the integrity of your data. This often involves using context managers (async with) to handle sessions and transactions.

- **Error handling**: Asynchronous code can make error handling trickier, especially with multiple concurrent tasks. Use try-except blocks to catch and handle exceptions appropriately.

- **Testing**: Testing asynchronous code requires some additional considerations. Make sure your test framework supports asynchronous tests and use `async` and `await` in your test cases as needed.

By understanding and applying these concepts, you can build applications that are not only robust but also capable of performing optimally under various load conditions. This knowledge is a valuable addition to your skillset as a modern web developer working with FastAPI.

See also

An overview of the concurrency use of the `asyncio` library in FastAPI can be found on the documentation page:

- *FastAPI Concurrency*: `https://fastapi.tiangolo.com/async/`

To integrate `async/await` syntax with **SQLAlchemy**, you can have a look at documentation support:

- *SQLAlchemy Asyncio*: `https://docs.sqlalchemy.org/en/20/orm/extensions/asyncio.html`

Chapter 6, *Integrating FastAPI with SQL Databases*, will focus on SQL database interactions. Here, you can find examples of integrating `asyncio` with `sqlalchemy`.

To integrate `asyncio` with **MongoDB**, you have to use a dedicated package called `motor`, which is built on top of `pymongo`:

- *Motor asynchronous driver*: `https://motor.readthedocs.io/en/stable/`

In *Chapter 7*, *Integrating FastAPI with NoSQL Databases*, you will find examples of motor integration with FastAPI.

Securing sensitive data and best practices

In the realm of web development, the security of sensitive data is paramount.

This recipe is a checklist of best practices for securing sensitive data in your FastAPI applications.

Getting ready

First and foremost, it's crucial to understand the types of data that need protection. *Sensitive data* can include anything from passwords and tokens to personal user details. Handling such data requires careful consideration and adherence to security best practices.

Understanding the types of data that require protection sets the foundation for implementing robust security measures, such as leveraging environment variables for sensitive configurations, a key aspect of data security in app development.

Instead of hardcoding these values in your source code, they should be stored in environment variables, which can be accessed securely within your application. This approach not only enhances security but also makes your application more flexible and easier to configure across different environments.

Another important practice is encrypting sensitive data, particularly passwords. FastAPI doesn't handle encryption directly, but you can use libraries such as `bcrypt` or `passlib` to hash and verify passwords securely.

This recipe will provide a checklist of good practices to apply to secure sensitive data.

How to do it...

Securely handling data in FastAPI involves more than just encryption; it encompasses a range of practices that are designed to protect data throughout its life cycle in your application.

Here is a list of good practices to apply when securing your application.

- **Validation and sanitization**: Use the Pydantic model to validate and sanitize incoming data, as shown in the *Working with data validation and serialization* recipe. Ensure the data conforms to expected formats and values, reducing the risk of injection attacks or malformed data causing issues.

  Be cautious with data that will be output to users or logs. Sensitive information should be redacted or anonymized to prevent accidental disclosure.

- **Access control**: Implement robust access control mechanisms to ensure that users can only access the data they are entitled to. This can involve **role-based access control** (**RBAC**), permission checks and properly managing user authentication. You will discover more about this in the *Setting up RBAC* recipe in *Chapter 4, Authentication and Authorization*.

- **Secure communication**: Use HTTPS to encrypt data in transit. This prevents attackers from intercepting sensitive data that's sent to or received from your application.

- **Database security**: Ensure that your database is securely configured. Use secure connections, avoid exposing database ports publicly, and apply the principle of least privilege to database access.

- **Regular updates**: Keep your dependencies, including FastAPI and its underlying libraries, up to date. This helps protect your application from vulnerabilities discovered in older versions of the software.

Some of them will be covered in detail throughout this book.

There's more...

Managing sensitive data extends beyond immediate security practices and involves considerations for data storage, transmission, and even deletion.

Here's a checklist of more general practices so that you can secure your data, regardless of whatever code you are writing:

- **Data storage**: Store sensitive data only when necessary. If you don't need to store data such as credit card numbers or personal identification numbers, then don't. When storage is necessary, ensure it is encrypted and that access is tightly controlled.

- **Data transmission**: Be cautious when transmitting sensitive data. Use secure APIs and ensure that any external services you interact with also follow security best practices.

- **Data retention and deletion**: Have clear policies on data retention and deletion. When data is no longer needed, ensure it is deleted securely, leaving no trace in backups or logs.

- **Monitoring and logging**: Implement monitoring to detect unusual access patterns or potential breaches. However, be careful with what you log. Avoid logging sensitive data and ensure that logs are stored securely and are only accessible to authorized personnel.

By applying these practices, you can significantly enhance the security posture of your applications, protecting both your users and your organization from potential data breaches and ensuring compliance with data protection regulations. As a developer, understanding and implementing data security is not just a skill but a responsibility in today's digital landscape. In the next chapter, we will learn how to build an entire RESTful API with FastAPI.

3

Building RESTful APIs with FastAPI

In this chapter, we delve into the essentials of building **RESTful APIs**. RESTful APIs are the backbone of web services, enabling applications to communicate and exchange data efficiently.

You will build a RESTful API for a Task Manager application. The application will interact with a CSV file, although the typical approach for such applications would be to use a database such as SQL or NoSQL. This approach is unconventional and not recommended for most scenarios due to scalability and performance limitations. However, in certain contexts, particularly in legacy systems or when dealing with large volumes of structured data files, managing data through CSV can be a practical solution.

Our Task Manager API will allow users to **create, read, update, and delete** (CRUD) tasks, each represented as a record in a CSV file. This example will provide insights into handling data in non-standard formats within FastAPI.

We will see how to test the API's endpoint. As your API grows, managing complex queries and filtering becomes essential. We'll explore techniques to implement advanced query capabilities, enhancing the usability and flexibility of your API.

Furthermore, we'll tackle the important aspect of versioning your API. Versioning is key to evolving your API over time without breaking existing clients. You'll learn strategies to manage API versions, ensuring backward compatibility and smooth transitions for users.

Lastly, we'll cover securing API with OAuth2, an industry-standard protocol for authorization. Security is paramount in API development, and you'll gain practical experience in implementing authentication and protecting your endpoints.

In this chapter, we're going to cover the following recipes:

- Creating CRUD operations
- Creating RESTful endpoints
- Testing your RESTful API
- Handling complex queries and filtering
- Versioning your API
- Securing your API with OAuth2
- Documenting your API with Swagger and Redoc

Technical requirements

To fully engage with this chapter in our *FastAPI Cookbook* and effectively build RESTful APIs, you'll need to have the following technologies and tools installed and configured:

- **Python**: Make sure you have a Python version higher than 3.9 in your environment.
- **FastAPI**: This should be installed with all required dependencies. If you haven't done it from the previous chapters, you can do so simply from your terminal with the following command:

  ```
  $ pip install fastapi[all]
  ```

- **Pytest**: You can install this framework by running the following:

  ```
  $ pip install pytest
  ```

Note that it can be very useful to already have some knowledge of the Pytest framework to better follow the *Testing your RESTful API* recipe.

The code used in the chapter is available on GitHub at the address: `https://github.com/PacktPublishing/FastAPI-Cookbook/tree/main/Chapter03`.

Feel free to follow along or consult it in case you get stuck.

Creating CRUD operations

This recipe will show you how to make the basic CRUD operations work with the CSV file that acts as a database.

We will begin by making a draft for a simple list of tasks in CSV format and we will put the operations in a separate Python module. By the end of the recipe, you will have all the operations ready to be used by the API's endpoints.

How to do it...

Let's start by creating a project root directory called `task_manager_app` to host our code base for our application:

1. Move into the root project folder and create a `tasks.csv` file, which we will use as a database and put a few tasks inside:

   ```
   id,title,description,status
   1,Task One,Description One,Incomplete
   2,Task Two,Description Two,Ongoing
   ```

2. Then, create a file called `models.py`, containing the Pydantic models that we will use internally for the code. It will look like the following:

   ```python
   from pydantic import BaseModel


   class Task(BaseModel):
       title: str
       description: str
       status: str


   class TaskWithID(Task):
       id: int
   ```

   We created two separate classes for task objects because `id` won't be used all along the code.

3. In a new file called `operations.py`, we will define the function that interacts with our database.

   We can start creating the CRUD operation

   Create a function to retrieve all the tasks from a `.csv` file:

   ```python
   import csv
   from typing import Optional

   from models import Task, TaskWithID

   DATABASE_FILENAME = "tasks.csv"

   column_fields = [
       "id", "title", "description", "status"
   ]

   def read_all_tasks() -> list[TaskWithID]:
   ```

```
with open(DATABASE_FILENAME) as csvfile:
    reader = csv.DictReader(
        csvfile,
    )
    return [TaskWithID(**row) for row in reader]
```

4. Now, we need to create a function to read a specific task based on `id`:

```
def read_task(task_id) -> Optional[TaskWithID]:
    with open(DATABASE_FILENAME) as csvfile:
        reader = csv.DictReader(
            csvfile,
        )
        for row in reader:
            if int(row["id"]) == task_id:
                return TaskWithID(**row)
```

5. To write a task, we need a strategy to assign a new `id` to the task that will written into the database.

    A good strategy can be to implement a logic based on the IDs already present in the database, then write the task into our CSV file, and group both operations into a new function. We can split the create task operation into three functions.

    First, let's create the function that retrieves the new ID based on the existing ones in the database:

```
def get_next_id():
    try:
        with open(DATABASE_FILENAME, "r") as csvfile:
            reader = csv.DictReader(csvfile)
            max_id = max(
                int(row["id"]) for row in reader
            )
            return max_id + 1
    except (FileNotFoundError, ValueError):
        return 1
```

Then, we define the function to write the task with the ID in the CSV file:

```
def write_task_into_csv(
    task: TaskWithID
):
    with open(
        DATABASE_FILENAME, mode="a", newline=""
    ) as file:
```

```
        writer = csv.DictWriter(
            file,
            fieldnames=column_fields,
        )
        writer.writerow(task.model_dump())
```

After that, we can leverage these last two functions to define the function that creates the task:

```
def create_task(
    task: Task
) -> TaskWithID:
    id = get_next_id()
    task_with_id = TaskWithID(
        id=id, **task.model_dump()
    )
    write_task_into_csv(task_with_id)
    return task_with_id
```

6.   Then, let's create the function to modify the task:

```
def modify_task(
    id: int, task: dict
) -> Optional[TaskWithID]:
    updated_task: Optional[TaskWithID] = None
    tasks = read_all_tasks()
    for number, task_ in enumerate(tasks):
        if task_.id == id:
            tasks[number] = (
                updated_task
            ) = task_.model_copy(update=task)
    with open(
        DATABASE_FILENAME, mode="w", newline=""
    ) as csvfile:  # rewrite the file
        writer = csv.DictWriter(
            csvfile,
            fieldnames=column_fields,
        )
        writer.writeheader()
        for task in tasks:
            writer.writerow(task.model_dump())
    if updated_task:
        return updated_task
```

7. Finally, let's create the function to remove the task with a specific `id`:

```python
def remove_task(id: int) -> bool:
    deleted_task: Optional[Task] = None
    tasks = read_all_tasks()
    with open(
        DATABASE_FILENAME, mode="w", newline=""
    ) as csvfile:  # rewrite the file
        writer = csv.DictWriter(
            csvfile,
            fieldnames=column_fields,
        )
        writer.writeheader()
        for task in tasks:
            if task.id == id:
                deleted_task = task
                continue
            writer.writerow(task.model_dump())
    if deleted_task:
        dict_task_without_id = (
            deleted_task.model_dump()
        )
        del dict_task_without_id["id"]
        return Task(**dict_task_wihtout_id)
```

You've just created the basic CRUD operations. We are now ready to expose those operations through the API endpoints.

How it works...

The structure of your API is fundamental in RESTful design. It involves defining endpoints (URIs) and associating them with HTTP methods to perform the desired operations.

In our Task Management system, we'll create endpoints to handle tasks, mirroring common CRUD operations. Here's an overview:

- `List Tasks` (`GET /tasks`) retrieves a list of all tasks
- `Retrieve Task` (`GET /tasks/{task_id}`) gets details of a specific task
- `Create Task` (`POST /task`) adds a new task
- `Update Task` (`PUT /tasks/{task_id}`) modifies an existing task
- `Delete Task` (`DELETE /tasks/{task_id}`) removes a task

Each endpoint represents a specific function in the API, clearly defined and purpose driven. FastAPI's routing system allows us to map these operations to Python functions easily.

> **Exercise**
>
> Try to write your unit tests for each one of the CRUD operations. If you follow along with the GitHub repository, you can find the tests in the `Chapter03/task_manager_rest_api/test_operations.py` file.

Creating RESTful Endpoints

Now, we will create the routes to expose each of the CRUD operations with a specific endpoint. In this recipe, we will see how FastAPI leverages Python type annotations to define expected request and response data types, streamlining the process of validation and serializing data.

Getting ready...

Before starting the recipe, make sure you know how to set up your local environment and create a basic FastAPI server. You can review it in the *Creating a new FastAPI project* and *Understanding FastAPI basics* recipes in *Chapter 1*, *First Steps with FastAPI*.

Also, we will use the CRUD operations created in the previous recipe.

How to do it...

Let's create a `main.py` file in the project root folder to code the server with the endpoints. FastAPI simplifies the implementation of different HTTP methods, aligning them with the corresponding CRUD operations.

Let's now write the endpoints for each operation:

1. Create the server with the endpoint to list all the tasks by using the `read_all_tasks` operation:

    ```
    from fastapi import FastAPI, HTTPException

    from models import (
        Task,
        TaskWithID,
    )
    from operations import read_all_tasks


    app = FastAPI()
    ```

```python
@app.get("/tasks", response_model=list[TaskWithID])
def get_tasks():
    tasks = read_all_tasks()
    return tasks
```

2. Now, let's write the endpoint to read a specific task based on `id`:

```python
@app.get("/task/{task_id}")
def get_task(task_id: int):
    task = read_task(task_id)
    if not task:
        raise HTTPException(
            status_code=404, detail="task not found"
        )
    return task
```

3. The endpoint to add a task will be as follows:

```python
from operations import create_task


@app.post("/task", response_model=TaskWithID)
def add_task(task: Task):
    return create_task(task)
```

4. To update the task, we can modify each field (`description`, `status`, or `title`). To do this, we create a specific model to be used in the body called `UpdateTask`. The endpoint will look like this:

```python
from operations import modify_task


class UpdateTask(BaseModel):
    title: str | None = None
    description: str | None = None
    status: str | None = None


@app.put("/task/{task_id}", response_model=TaskWithID)
def update_task(
    task_id: int, task_update: UpdateTask
):
    modified = modify_task(
        task_id,
```

```
            task_update.model_dump(exclude_unset=True),
        )
        if not modified:
            raise HTTPException(
                status_code=404, detail="task not found"
            )

        return modified
```

5. Finally, here is the endpoint to delete a task:

```
from operations import remove_task

@app.delete("/task/{task_id}", response_model=Task)
def delete_task(task_id: int):
    removed_task = remove_task(task_id)
    if not removed_task:
        raise HTTPException(
            status_code=404, detail="task not found"
        )
    return removed_task
```

You have just implemented the operations to interact with the CSV file used as a database.

From a command terminal at the project root folder level, spin up the server with the uvicorn command:

```
$ uvicorn main:app
```

In the browser, go to http://localhost:8000/docs and you will see the endpoints of your RESTful API that you just made.

You can experiment by creating some tasks, then listing them, updating them, and deleting some of them directly with the interactive documentation.

Testing your RESTful API

Testing is a critical part of API development. In FastAPI, you can use various testing frameworks such as pytest to write tests for your API endpoints.

In this recipe, we are going to write unit tests for each of the endpoints we created earlier.

Getting ready...

If not done yet, ensure you have `pytest` installed in your environment by running:

```
$ pip install pytest
```

It's a good practice in testing to use a dedicated database to avoid interaction with the production one. To accomplish this, we will create a test fixture that generates the database before each test.

We will define this in a `conftest.py` module so that the fixture is applied to all tests under the project's root folder. Let's create the module in the project root folder and start by defining a list of test tasks and the name of the CSV file used for the tests:

```
TEST_DATABASE_FILE = "test_tasks.csv"

TEST_TASKS_CSV = [
    {
        "id": "1",
        "title": "Test Task One",
        "description": "Test Description One",
        "status": "Incomplete",
    },
    {
        "id": "2",
        "title": "Test Task Two",
        "description": "Test Description Two",
        "status": "Ongoing",
    },
]

TEST_TASKS = [
    {**task_json, "id": int(task_json["id"])}
    for task_json in TEST_TASKS_CSV
]
```

We can now create a fixture that will be used for all the tests. This fixture will set up the test database before each test function execution.

We can achieve this by passing the `autouse=True` argument to the `pytest.fixture` decorator, which indicates that the feature will run before every single test:

```
import csv
import os
from pathlib import Path
```

```
from unittest.mock import patch

import pytest

@pytest.fixture(autouse=True)
def create_test_database():
    database_file_location = str(
        Path(__file__).parent / TEST_DATABASE_FILE
    )
    with patch(
        "operations.DATABASE_FILENAME",
        database_file_location,
    ) as csv_test:
        with open(
            database_file_location, mode="w", newline=""
        ) as csvfile:
            writer = csv.DictWriter(
                csvfile,
                fieldnames=[
                    "id",
                    "title",
                    "description",
                    "status",
                ],
            )
            writer.writeheader()
            writer.writerows(TEST_TASKS_CSV)
            print("")
        yield csv_test
        os.remove(database_file_location)
```

Since the fixture is defined in a `conftest.py` module, each test module will automatically import it.

Now, we can proceed with creating the actual unit test functions for the endpoints created in the previous recipe.

How to do it...

To test the endpoints, FastAPI provides a specific `TestClient` class that allows the testing of the endpoints without running the server.

In a new module called `test_main.py`, let's define our test client:

```
from main import app
from fastapi.testclient import TestClient

client = TestClient(app)
```

We can create the tests for each endpoint as follows.

1. Let's start with the GET /tasks endpoint, which lists all the tasks in the database:

    ```
    from conftest import TEST_TASKS

    def test_endpoint_read_all_tasks():
        response = client.get("/tasks")
        assert response.status_code == 200
        assert response.json() == TEST_TASKS
    ```

    We are asserting the response's status code and the json body.

2. As easy as that, we can go on by creating the test for GET /tasks/{task_id} to read a task with a specific id:

    ```
    def test_endpoint_get_task():
        response = client.get("/task/1")

        assert response.status_code == 200
        assert response.json() == TEST_TASKS[0]

        response = client.get("/task/5")

        assert response.status_code == 404
    ```

    Besides the 200 status code for an existing task, we also asserted the status code is equal to 404 when the task does not exist in the database.

3. In a similar way, we can test the POST /task endpoint to add a new task into the database by asserting the new assigned id for the task:

    ```
    from operations import read_all_tasks

    def test_endpoint_create_task():
        task = {
            "title": "To Define",
    ```

```
        "description": "will be done",
        "status": "Ready",
    }
    response = client.post("/task", json=task)

    assert response.status_code == 200
    assert response.json() == {**task, "id": 3}
    assert len(read_all_tasks()) == 3
```

4. The test for the PUT /tasks/{task_id} endpoint to modify a task will then be the following:

```
from operations import read_task


def test_endpoint_modify_task():
    updated_fields = {"status": "Finished"}
    response = client.put(
        "/task/2", json=updated_fields
    )

    assert response.status_code == 200
    assert response.json() == {
        *TEST_TASKS[1],
        *updated_fields,
    }

    response = client.put(
        "/task/3", json=updated_fields
    )

    assert response.status_code == 404
```

5. Finally, we test the DELETE /tasks/{task_id} endpoint to delete a task:

```
def test_endpoint_delete_task():
    response = client.delete("/task/2")
    assert response.status_code == 200

    expected_response = TEST_TASKS[1]
    del expected_response["id"]

    assert response.json() == expected_response
    assert read_task(2) is None
```

You've just written all the unit tests for each of the API endpoints.

You can now run the tests from the project root folder by running in the terminal, or with the GUI support of your favorite editor:

```
$ pytest .
```

Pytest will collect all the tests and run them. If everything is correctly done, you will see a message that says you got a 100% score in the output of the console if you have written the tests correctly.

See also

You can check test fixtures in the Pytest documentation:

- *Pytest Fixtures Reference*: `https://docs.pytest.org/en/7.1.x/reference/fixtures.html`

You can dig into FastAPI testing tools and the `TestClient` API in the official documentation:

- *FastAPI Testing*: `https://fastapi.tiangolo.com/tutorial/testing/`
- *FastAPI TestClient*: `https://fastapi.tiangolo.com/reference/testclient/`

Handling complex queries and filtering

In any RESTful API, providing the functionality to filter data based on certain criteria is essential. In this recipe, we'll enhance our Task Manager API to allow users to filter tasks based on different parameters and create a search endpoint.

Getting ready...

The filtering functionality will be implemented in the existing `GET /tasks` endpoint to show how to overcharge an endpoint, while the search functionality will be shown on a brand-new endpoint. Make sure you have at least the CRUD operations already in place before continuing.

How to do it...

We will start by overcharging `GET /tasks` endpoint with filters. We modify the endpoint to accept two query parameters: `status` and `title`.

The endpoint will then look like the following:

```python
@app.get("/tasks", response_model=list[TaskWithID])
def get_tasks(
    status: Optional[str] = None,
    title: Optional[str] = None,
):
    tasks = read_all_tasks()
    if status:
        tasks = [
            task
            for task in tasks
            if task.status == status
        ]
    if title:
        tasks = [
            task for task in tasks if task.title == title
        ]
    return tasks
```

The two parameters can be optionally specified to filter the tasks that match their value.

Next, we implement a search functionality. Beyond basic filtering, implementing a search functionality can significantly improve the usability of an API. We'll add a search feature that allows users to find tasks based on a keyword present in the title or description in a new endpoint:

```python
@app.get("/tasks/search", response_model=list[TaskWithID])
def search_tasks(keyword: str):
    tasks = read_all_tasks()
    filtered_tasks = [
        task
        for task in tasks
        if keyword.lower()
        in (task.title + task.description).lower()
    ]
    return filtered_tasks
```

In the `search_tasks` endpoint, the function filters tasks to include only those where the keyword appears in either the title or the description.

To start the server as usual, run this command from the command line:

```
$ uvicorn main:app
```

Then, go to the interactive documentation address at `http://localhost:8000/docs`, and you will see the new endpoint we've just made.

Play around by specifying some keywords that could be in the title or the description of one of your tasks.

Versioning your API

API versioning is essential in maintaining and evolving web services without disrupting the existing users. It allows developers to introduce changes, improvements, or even breaking changes while providing backward compatibility. In this recipe, we will implement versioning in our Task Manager API.

Getting ready...

To follow the recipe, you will need to have endpoints already defined. If you don't have them, you can first check the *Creating RESTful endpoints* recipe.

How to do it...

There are several strategies for API versioning. We will use the most common approach, URL path versioning, for our API.

Let's consider that we want to improve the task information by adding a new `str` field called `priority` that is set to `"lower"` by default. Let's do it through the following steps.

1. Let's create a `TaskV2` object class in the `models.py` module:

    ```python
    from typing import Optional

    class TaskV2(BaseModel):
        title: str
        description: str
        status: str
        priority: str | None = "lower"


    class TaskV2WithID(TaskV2):
        id: int
    ```

2. In the `operations.py` module, let's create a new function called `read_all_tasks_v2`, which reads all the tasks, and add the `priority` field:

```
from models import TaskV2WIthID

def read_all_tasks_v2() -> list[TaskV2WIthID]:
    with open(DATABASE_FILENAME) as csvfile:
        reader = csv.DictReader(
            csvfile,
        )
        return [TaskV2WIthID(**row) for row in reader]
```

3. We have now all we need to create version two of `read_all_tasks` function. We will do this in the `main.py` module:

```
from models import TaskV2WithID
@app.get(
    "/v2/tasks",
    response_model=list[TaskV2WithID]
)
def get_tasks_v2():
    tasks = read_all_tasks_v2()
    return tasks
```

You've just created version two of the endpoint. In this way, you can develop and improve your API with several versions of your endpoint.

To test it, let's modify our `tasks.csv` file by manually adding the new field to test the new endpoint:

```
id,title,description,status,priority
1,Task One,Description One,Incomplete
2,Task Two,Description Two,Ongoing,higher
```

Start the server once more from the command line:

```
$ uvicorn main:app
```

Now, the interactive documentation at `http://localhost:8000/docs` will show the new GET `/v2/tasks` endpoint to list all the tasks in version 2 mode.

Check that the endpoint lists the tasks with the new `priority` field and that the old GET `/tasks` is still working as expected.

> **Exercise**
> You might have noticed that using a CSV file as a database might not be the most reliable solution. If the process crashes during an update or removal, you can lose all of the data. So, improve the API with a newer version of the endpoints that use operational functions that interact with an SQLite database.

There's more...

When you version an API, you are essentially providing a way to differentiate between different releases or versions of your API, allowing clients to choose which version they want to interact with.

Besides the URL-based approach that we used in the recipe, there are other common approaches to API versioning, such as the following:

- **Query parameter versioning**: Version information is passed as a query parameter in the API request. For example, see the following:

  ```
  https://api.example.com/resource?version=1
  ```

  This method keeps the base URL uniform across versions.

- **Header versioning**: The version is specified in a custom header of the HTTP request:

  ```
  GET /resource HTTP/1.1
  Host: api.example.com
  X-API-Version: 1
  ```

  This keeps the URL clean but requires clients to explicitly set the version in their requests.

- **Consumer-based versioning**: This strategy allows customers to choose the version they need. The version available at their first interaction is saved with their details and used in all future interactions unless they make changes.

Furthermore, it can be relevant to use **semantic versioning** where version numbers follow the semantic versioning format (MAJOR.MINOR.PATCH). Changes in the MAJOR version indicate incompatible API changes, while MINOR and PATCH versions indicate backward-compatible changes.

Versioning allows API providers to introduce changes (such as adding new features, modifying existing behavior, or deprecating endpoints and sunset policies) without breaking existing client integrations.

It also gives consumers control over when and how they adopt new versions, minimizing disruptions and maintaining stability in the API ecosystem.

See also

You can have a look at an interesting article from the Postman blog on API versioning strategies:

- *Postman Blog API Versioning*: `https://www.postman.com/api-platform/api-versioning/`

Securing your API with OAuth2

In web applications, securing endpoints from unauthorized users is crucial. **OAuth2** is a common authorization framework that enables applications to be accessed by user accounts with restricted permissions. It works by issuing tokens instead of credentials. This recipe will show how to use OAuth2 in our Task Manager API to protect endpoints.

Getting ready...

FastAPI provides support for OAuth2 with a password, including the use of external tokens. Data compliance regulations require that passwords are not stored in plain text. Instead, a usual method is to store the outcome of the hashing operation, which changes the plain text into a string that is not readable by humans and cannot be reversed.

> **Important note**
> With the only purpose of showing the functionality, we will fake the hashing mechanism as well the token creation with trivial ones. For obvious security reasons, do not use it in a production environment.

How to do it...

Let's start by creating a `security.py` module in the project root folder where we are going to implement all tools used to secure our service. Then let's create a secured endpoint as follows.

1. First, let's create a dictionary containing a list of users with their usernames and passwords:

```
fake_users_db = {
    "johndoe": {
        "username": "johndoe",
        "hashed_password": "hashedsecret",
    },
    "janedoe": {
        "username": "janedoe",
        "hashed_password": "hashedsecret2",
    },
}
```

2. Passwords should not be stored in plain text, but encrypted or hashed. To demonstrate the
    feature, we fake the hashing mechanism by inserting `"hashed"` before the password string:

```python
def fakely_hash_password(password: str):
    return f"hashed{password}"
```

3. Let's create the classes to handle the users and a function to retrieve the user from the `dict`
    database we created:

```python
class User(BaseModel):
    username: str


class UserInDB(User):
    hashed_password: str


def get_user(db, username: str):
    if username in db:
        user_dict = db[username]
        return UserInDB(**user_dict)
```

4. Using a similar logic to what we've just used for hashing, let's make a fake token generator and
    a fake token resolver:

```python
def fake_token_generator(user: UserInDB) -> str:
    # This doesn't provide any security at all
    return f"tokenized{user.username}"


def fake_token_resolver(
    token: str
) -> UserInDB | None:
    if token.startswith("tokenized"):
        user_id = token.removeprefix("tokenized")
        user = get_user(fake_users_db, user_id)
        return user
```

5. Now, let's create a function to retrieve the user from the token. To this, we will make use of the Depends class to use dependency injection provided by FastAPI (see https://fastapi. tiangolo.com/tutorial/dependencies/), with the OAuthPasswordBearer class to handle the token:

```
from fastapi import Depends, HTTPException, status

oauth2_scheme = OAuth2PasswordBearer(tokenUrl="token")


def get_user_from_token(
    token: str = Depends(oauth2_scheme),
) -> UserInDB:
    user = fake_token_resolver(token)
    if not user:
        raise HTTPException(
            status_code=status.HTTP_401_UNAUTHORIZED,
            detail=(
                "Invalid authentication credentials"
            ),
            headers={"WWW-Authenticate": "Bearer"},
        )
    return user
```

oauth2scheme contains the /token URL endpoint that will be used by the interactive documentation to authenticate the browser.

> **Important note**
> We have used a dependency injection to retrieve the token from the get_user_token function with the fastapi.Depends object. A dependency injection pattern is not native to the Python language and it is strictly related to the FastAPI framework. In *Chapter 8, Advanced Features and Best Practices*, you will find a dedicated recipe about that called *Implementing dependency injection*.

6. Let's create the endpoint in the main.py module:

```
from fastapi import Depends, HTTPException
from fastapi.security import OAuth2PasswordRequestForm
from security import (
    UserInDB,
    fake_token_generator,
    fakely_hash_password,
    fake_users_db
```

```
)

@app.post("/token")
async def login(
    form_data: OAuth2PasswordRequestForm = Depends(),
):
    user_dict = fake_users_db.get(form_data.username)
    if not user_dict:
        raise HTTPException(
            status_code=400,
            detail="Incorrect username or password",
        )
    user = UserInDB(**user_dict)
    hashed_password = fakely_hash_password(
        form_data.password
    )
    if not hashed_password == user.hashed_password:
        raise HTTPException(
            status_code=400,
            detail="Incorrect username or password",
        )

    token = fake_token_generator(user)

    return {
        "access_token": token,
        "token_type": "bearer"
    }
```

We now have all we need to create a secured endpoint with OAuth2 authentication.

7. The endpoint we are going to create will return information about the current user from the token provided. If the token does not have authorization, it will return a 400 exception:

```
from security import get_user_from_token

@app.get("/users/me", response_model=User)
def read_users_me(
    current_user: User = Depends(get_user_from_token),
):
    return current_user
```

The endpoint we just created will be reachable only by allowed users.

Let's now test our secured endpoint. From the command line terminal at the project root folder level, spin up the server by running:

```
$ uvicorn main:app
```

Then, open the browser, go to `http://localhost:8000/docs`, and you will notice the new `token` and `users/me` endpoints in the interactive documentation.

You might notice a little padlock icon on the `users/me` endpoint. If you click on it, you will see a form window that allows you to get the token and store it directly in your browser, so you don't have to provide it each time you call the secured endpoint.

> **Exercise**
>
> You've just learned how to create a secured endpoint for your RESTful API. Now, try to secure some of the endpoints you created in the previous recipes.

There's more...

With OAuth2, we can define a **scope** parameter, which is used to specify the level of access that an access token grants to a client application when it is used to access a protected resource. Scopes can be used to define what actions or resources the client application is allowed to perform or access on behalf of the user.

When a client requests authorization from the resource owner (user), it includes one or more scopes in the authorization request. In FastAPI, these scopes are represented as `dict`, where keys represent the scope's name and the value is a description.

The authorization server then uses these scopes to determine the appropriate access controls and permissions to grant to the client application when issuing an access token.

It is not the purpose of this recipe to go into the details of implementing OAuth2 scopes in FastAPI. However, you can find practical examples on the official documentation page at the link: `https://fastapi.tiangolo.com/advanced/security/oauth2-scopes/`.

See also

You can check on how FastAPI integrates OAuth2 at the following link:

- *Simple OAuth2 with Password and Bearer*: `https://fastapi.tiangolo.com/tutorial/security/simple-oauth2/`

Also, you can find more on dependency injection in FastAPI on the official documentation page:

- *Dependencies*: `https://fastapi.tiangolo.com/tutorial/dependencies/`

Documenting your API with Swagger and Redoc

FastAPI automatically generates documentation for your API using **Swagger UI** and **Redoc**, when spinning the server.

This documentation is derived from your route functions and Pydantic models, making it incredibly beneficial for both development and consumption by frontend teams or API consumers.

In this recipe, we will see how to customize the documentation's specific needs.

Getting ready...

By default, FastAPI provides two documentation interfaces:

- **Swagger UI**: Accessible at /docs endpoint (e.g., http://127.0.0.1:8000/docs)
- **Redoc**: Accessible at /redoc endpoint (e.g., http://127.0.0.1:8000/redoc)

These interfaces offer dynamic documentation where users can see and test the API endpoints and their details. However, both pieces of documentation can be modified.

How to do it...

FastAPI allows the customization of Swagger UI. You can add metadata, customize the look, and add additional documentation through the FastAPI class parameters.

You can enhance your API documentation by providing additional metadata such as title, description, and version to the app object in the main.py module:

```
app = FastAPI(
    title="Task Manager API",
    description="This is a task management API",
    version="0.1.0",
)
```

This metadata will appear in both Swagger UI and Redoc documentation.

You can push things further by completely customizing your Swagger UI in case you need to expose it to a third user under certain conditions.

Let's try to hide the /token endpoint from the documentation.

In this case, you can use the `utils`, module provided by FastAPI to retrieve the OpenAPI schema of the Swagger UI in a `dict` object as follows:

```
from fastapi.openapi.utils import get_openapi

def custom_openapi():
    if app.openapi_schema:
        return app.openapi_schema
    openapi_schema = get_openapi(
        title="Customized Title",
        version="2.0.0",
        description="This is a custom OpenAPI schema",
        routes=app.routes,
    )
    del openapi_schema["paths"]["/token"]
    app.openapi_schema = openapi_schema
    return app.openapi_schema


app = FastAPI(
    title="Task Manager API",
    description="This is a task management API",
    version="0.1.0",
)

app.openapi = custom_openapi
```

That's all you need to customize your API documentation.

If you spin up the server with the `uvicorn main:app` command and go to one of the two documentation pages, the `/token` endpoint won't appear anymore.

You are now able to customize your API documentation to elevate the way you present it to your customers.

See also

You find out more about FastAPI generation for metadata, features, and OpenAPI integration on the official documentation pages:

- *Metadata and Docs URLs*: https://fastapi.tiangolo.com/tutorial/metadata/
- *FastAPI Features*: https://fastapi.tiangolo.com/features/
- *Extending OpenAPI*: https://fastapi.tiangolo.com/how-to/extending-openapi/

4

Authentication and Authorization

In this chapter of our *FastAPI Cookbook*, we will delve into the critical realms of authentication and authorization, laying the foundation to secure your web applications against unauthorized access.

As we navigate through this chapter, you'll embark on a practical journey to implement a comprehensive security model in your FastAPI applications. From the basics of user registration and authentication to the integration of sophisticated **OAuth2** protocols with **JSON Web Token** (**JWT**) for enhanced security, this chapter covers it all.

We will create the essential components of **software as a service** (**SaaS**) to help you learn practically how to establish user registration systems, verify users, and handle sessions efficiently. We'll also show you how to apply **role-based access control** (**RBAC**) to adjust user permissions and protect API endpoints with API key authentication. The incorporation of third-party authentication using external login services, such as GitHub, will demonstrate how to leverage existing platforms for user authentication, simplifying the login process for your users.

Furthermore, you'll add an extra layer of security by implementing **multi-factor authentication** (**MFA**), ensuring that your application's security is robust against various attack vectors.

In this chapter, we're going to cover the following recipes:

- Setting up user registration
- Working with OAuth2 and JWT for authentication
- Setting up RBAC
- Using third-party authentication
- Implementing MFA
- Handling API key authentication
- Handling session cookies and logout functionality

Technical requirements

To dive into the chapter and follow along with recipes on authentication and authorization, ensure your setup includes the following essentials:

- **Python**: Install a Python version higher than 3.9 in your environment.

- **FastAPI**: This should be installed with all required dependencies. If you didn't do so during the previous chapters, you can simply do it from your terminal:

    ```
    $ pip install fastapi[all]
    ```

The code used in the chapter is hosted on GitHub at `https://github.com/PacktPublishing/FastAPI-Cookbook/tree/main/Chapter04`.

Setting up a virtual environment for the project within the project root folder is also recommended to manage dependencies efficiently and maintain project isolation. Within your virtual environment, you can install all the dependencies at once by using the `requirements.txt` file, provided in the GitHub repository in the project folder:

```
pip install -r requirements.txt
```

Since the interactive Swagger documentation is limited at the time of writing, a basic mastering of **Postman** or any other testing API is beneficial to test our API.

Now that we have this ready, we can begin preparing our recipes.

Setting up user registration

User registration is the first step in securing your FastAPI application. It involves collecting user details and storing them securely. Here's how you can set up a basic user registration system. The recipe will show you how to set up a FastAPI application's registration system.

Getting ready

We will start by storing users in an SQL database. Let's create a project root folder called `saas_app`, containing the code base.

To store user passwords, we will use an external package to hash plain text with the **bcrypt** algorithm. The hashing function transforms a text string into a unique and irreversible output, allowing for secure storage of sensitive data such as passwords. You can find more details at `https://en.wikipedia.org/wiki/Hash_function`.

If you haven't installed packages from `requirements.txt` from the GitHub repository of the chapter under the `saas_app` project folder,, you can install the `passlib` package with `bycrypt` by running the following:

```
$ pip install passlib[bcrypt]
```

You will also need to install a version of `sqlalchemy` higher than 2.0.0 to follow along with the code in the GitHub repository:

```
$ pip install sqlalchemy>=2.0.0
```

Our environment is now ready to implement the user registration in our SaaS.

How to do it...

Before starting the implementation, we need to set up the database to store our users.

We need to set up a **SQLite** database with `sqlalchemy` for the application to store user credentials.

You need to do the following:

- Set up a `User` class to map the users table in the SQL database. The table should contain the `id`, `username`, `email`, and `hashed_password` fields.
- Establish the connection between the application and the database.

First let's create our project root folder called `saas_app`. Then you can refer to the *Setting up SQL databases* recipe in *Chapter 2, Working with Data*, or copy the `database.py` and `db_connection.py` modules from the GitHub repository under your root folder.

With the database session set up, let's define the function that adds a user.

Let's make it into a dedicated module called `operations.py`, in which we will define all the support functions used by the API endpoints.

The function will use a password context object from the `bcrypt` package to hash plain text passwords. We can define it as follows:

```
from passlib.context import CryptContext

pwd_context = CryptContext(
    schemes=["bcrypt"], deprecated="auto"
)
```

We can then define the `add_user` function, which inserts a new user into the database with the hashed password, according to most of data compliance regulations:

```
from sqlalchemy.exc import IntegrityError
from sqlalchemy.orm import Session

from models import User
```

```
def add_user(
    session: Session,
    username: str,
    password: str,
    email: str,
) -> User | None:
    hashed_password = pwd_context.hash(password)
    db_user = User(
        username=username,
        email=email,
        hashed_password=hashed_password,
    )
    session.add(db_user)
    try:
        session.commit()
        session.refresh(db_user)
    except IntegrityError:
        session.rollback()
        return
    return db_user
```

`InegrityError` will take into account the attempt to add a username or email that already exists.

We now have to define our endpoint, but first, we need to set up our server and initialize the database connection. We can do it in the `main.py` module, as follows:

```
from contextlib import (
    asynccontextmanager,
)

from fastapi import FastAPI

from db_connection import get_engine

@asynccontextmanager
async def lifespan(app: FastAPI):
    Base.metadata.create_all(bind=get_engine())
    yield


app = FastAPI(
```

```
            title="Saas application", lifespan=lifespan
)
```

We use the lifespan parameter of the FastAPI object to instruct the server to sync our database class, User, with the database when it starts up.

In addition, we can create a separate module, responses.py, to keep the response classes used for different endpoints. Feel free to create your own or copy the one provided in the GitHub repository.

We can now write the suitable endpoint to sign up a user in the same main.py module:

```
from typing import Annotated

from sqlalchemy.orm import Session
from fastapi import Depends, HTTPException, status

from models import Base
from db_connection import get_session
from operations import add_user

@app.post(
    "/register/user",
    status_code=status.HTTP_201_CREATED,
    response_model=ResponseCreateUser,
    responses={
        status.HTTP_409_CONFLICT: {
            "description": "The user already exists"
        }
    },
)
def register(
    user: UserCreateBody,
    session: Session = Depends(get_session),
) -> dict[str, UserCreateResponse]:
    user = add_user(
        session=session, **user.model_dump()
    )
    if not user:
        raise HTTPException(
            status.HTTP_409_CONFLICT,
            "username or email already exists",
        )
    user_response = UserCreateResponse(
```

```
            username=user.username, email=user.email
        )
    return {
        "message": "user created",
        "user": user_response,
    }
```

We have just implemented a basic mechanism to register and store users in our SaaS database.

How it works...

The endpoint will accept a JSON body containing a username, email, and password.

If the username or email already exists, a 409 response will be returned, and user creation will be disallowed.

To test this, at the project root level, spin up the server by running the following:

```
$ uvicorn main:app
```

Then, connect with your browser at localhost:8000/docs and check the endpoint we just created in the Swagger documentation. Feel free to play around with it.

Exercise

Create proper tests for both the add_user function and the /register/user endpoint, such as the following:

```
def test_add_user_into_the_database(session):
    user = add_user(...
    # fill in the test


def test_endpoint_add_basic_user(client):
    response = client.post(
        "/register/user",
        json=
    # continue the test
```

You can arrange the test in any way that works best for you.

You can find a possible way of testing in the Chapter04/saas_app folder of the book's GitHub repository.

See also

The **bcrypt** library allows you to add several layers of security to your hashing functions, such as salt and additional keys. Feel free to have a look at the source code on GitHub:

- *Bcrypt GitHub Repository*: `https://github.com/pyca/bcrypt/`

Also, you can find some interesting examples of how to use it at the following:

- *Hashing Passwords in Python with Bcrypt*: `https://www.geeksforgeeks.org/hashing-passwords-in-python-with-bcrypt/`

Working with OAuth2 and JWT for authentication

In this recipe, we'll integrate OAuth2 with JWTs for secure user authentication in your application. This approach improves security by utilizing tokens instead of credentials, aligning with modern authentication standards.

Getting ready

Since we will use a specific library to manage JWT, ensure you have the necessary dependencies installed. If you haven't installed the packages from `requirements.txt`, run the following:

```
$ pip install python-jose[cryptography]
```

Also, we will use the users table used in the previous recipe, *Setting up user registration*. Make sure to have set it up before starting the recipe.

How to do it...

We can set up the JWT token integration through the following steps.

1. In a new module called `security.py`, let's define the authentication function for the user:

```python
from sqlalchemy.orm import Session
from models import User
from email_validator import (
    validate_email,
    EmailNotValidError,
)

from operations import pwd_context

def authenticate_user(
    session: Session,
```

```
        username_or_email: str,
        password: str,
) -> User | None:
        try:
            validate_email(username_or_email)
            query_filter = User.email
        except EmailNotValidError:
            query_filter = User.username
        user = (
            session.query(User)
            .filter(query_filter == username_or_email)
            .first()
        )
        if not user or not pwd_context.verify(
            password, user.hashed_password
        ):
            return
        return user
```

The function can validate the input based on either the username or email.

2. Let's define the functions to create and decode the access token in the same module (create_access_token and decode_access_token).

    To create the access token, we will need to specify a secret key, the algorithm used to generate it, and the expiration time, as follows:

    ```
    SECRET_KEY = "a_very_secret_key"
    ALGORITHM = "HS256"
    ACCESS_TOKEN_EXPIRE_MINUTES = 30
    ```

    Then, the create_access_token_function is as follows:

    ```
    from jose import jwt

    def create_access_token(data: dict) -> str:
        to_encode = data.copy()
        expire = datetime.utcnow() + timedelta(
            minutes=ACCESS_TOKEN_EXPIRE_MINUTES
        )
        to_encode.update({"exp": expire})
        encoded_jwt = jwt.encode(
            to_encode, SECRET_KEY, algorithm=ALGORITHM
        )
        return encoded_jwt
    ```

To decode the access token, we can use a support function, `get_user`, that returns the `User` object by the username. You can do it on your own in the `operations.py` module or take it from the GitHub repository.

The function to decode the token will be as follows:

```
from jose import JWTError


def decode_access_token(
    token: str, session: Session
) -> User | None:
    try:
        payload = jwt.decode(
            token, SECRET_KEY, algorithms=[ALGORITHM]
        )
        username: str = payload.get("sub")
    except JWTError:
        return
    if not username:
        return
    user = get_user(session, username)
    return user
```

3. We can now proceed to create the endpoint to retrieve the token in the same module, `security.py`, with the `APIRouter` class:

```
from fastapi import (
    APIRouter,
    Depends,
    HTTPException,
    status,
)

from fastapi.security import (
    OAuth2PasswordRequestForm,
)

router = APIRouter()

class Token(BaseModel):
    access_token: str
    token_type: str


@router.post(
```

```
        "/token",
        response_model=Token,
        responses=..., # document the responses
    )
    def get_user_access_token(
        form_data: OAuth2PasswordRequestForm = Depends(),
        session: Session = Depends(get_session),
    ):
        user = authenticate_user(
            session,
            form_data.username,
            form_data.password
        )
        if not user:
            raise HTTPException(
                status_code=status.HTTP_401_UNAUTHORIZED,
                detail="Incorrect username or password",
            )
        access_token = create_access_token(
            data={"sub": user.username}
        )
        return {
            "access_token": access_token,
            "token_type": "bearer",
        }
```

4. Then, we can now create an `OAuth2PasswordBearer` object for the `POST /token` endpoint to obtain the access token:

```
from fastapi.security import (
    OAuth2PasswordBearer,
)

oauth2_scheme = OAuth2PasswordBearer(tokenUrl="token")
```

5. Finally, we can create the /users/me endpoint that returns the credentials based on the token:

```
@router.get(
    "/users/me",
    responses=..., # document responses
)
def read_user_me(
    token: str = Depends(oauth2_scheme),
```

```
            session: Session = Depends(get_session),
    ):
        user = decode_access_token(token, session)
        if not user:
            raise HTTPException(
                status_code=status.HTTP_401_UNAUTHORIZED,
                detail="User not authorized",
            )
        return {
            "description": f"{user.username} authorized",
        }
```

6. Now, let's import those endpoints into the FastAPI server in main.py. Right after defining the FastAPI object, let's add the router, as follows:

    ```
    import security

    # rest of the code

    app.include_router(security.router)
    ```

We have just defined the authentication mechanism for our SaaS.

How it works...

Now, spin up the server by running the following code from the terminal at the project root folder level:

```
$ uvicorn main:app
```

Go to the Swagger documentation address in your browser (localhost:8000/docs) and you will see the new endpoints, POST /token and GET /users/me.

You need the token to call the second endpoint, which you can store in your browser automatically by clicking on the lock icon and filling out the form with your credentials.

You've made your SaaS application more secure by using OAuth2 with JWT, which help you guard your sensitive endpoints and make sure that only users who are logged in can use them. This arrangement gives you a reliable and safe way to verify users that works well for modern web applications.

See also

You can gain a better understanding of the OAuth2 framework by reading this article:

- *Introduction to OAuth2*: `https://www.digitalocean.com/community/tutorials/an-introduction-to-oauth-2`

Also, you can have a look at the protocol definition for JWTs at the following:

- *JWT IETF Document*: `https://datatracker.ietf.org/doc/html/rfc7519`

Setting up RBAC

RBAC is a method of regulating access to resources based on the roles of individual users within an organization. In this recipe, we'll implement RBAC in a FastAPI application to manage user permissions effectively.

Getting ready

Since we will expand our database to accommodate role definitions, make sure you have completed the *Setting up user registration* recipe before diving into this.

To set up access control, we first need to define a variety of roles that we can allocate to. Let's follow these steps to do it.

1. In the `module.py` module, we can define a new class called `Role` and add it as a new field of the `User` model that will be stored in the users table:

```python
from enum import Enum

class Role(str, Enum):
    basic = "basic"
    premium = "premium"


class User(Base):
    __tablename__ = "users"
# existing fields
    role: Mapped[Role] = mapped_column(
        default=Role.basic
    )
```

2. Then, in the `operations.py` module, we will modify the `add_user` function in `operations.py` to accept a parameter to define the user role; the default value will be the basic role:

```python
from models import Role


def add_user(
    session: Session,
    username: str,
    password: str,
    email: str,
    role: Role = Role.basic,
) -> User | None:
    hashed_password = pwd_context.hash(password)
    db_user = User(
        username=username,
        email=email,
        hashed_password=hashed_password,
        role=role,
    )
    # rest of the function
```

3. Let's create a new module called `premium_access.py` and define the endpoint through a new router to sign up a premium user, which will look a lot like the endpoint to sign up a basic user:

```python
@router.post(
    "/register/premium-user",
    status_code=status.HTTP_201_CREATED,
    response_model=ResponseCreateUser,
    responses=..., # document responses
)
def register_premium_user(
    user: UserCreateBody,
    session: Session = Depends(get_session),
):
    user = add_user(
        session=session,
         *user.model_dump(),
        role=Role.premium,
    )
    if not user:
        raise HTTPException(
```

```
                status.HTTP_409_CONFLICT,
                "username or email already exists",
            )
    user_response = UserCreate(
        username=user.username,
        email=user.email,
    )
    return {
        "message": "user created",
        "user": user_response,
    }
```

In the previous code snippet, the imports and the router definition are skipped, since they are similar to the ones used in other modules.

4. Let's add the router to our app class in the `main.py` module:

```
import security
import premium_access

# rest of the code

app.include_router(security.router)
app.include_router(premium_access.router)
```

We have now all the elements to implement RBAC in our SaaS application.

How to do it...

Let's create two endpoints, one accessible for all the users and one reserved only for premium user. Let's make the endpoints through the following steps.

1. First, let's create two helper functions, `get_current_user` and `get_premium_user`, to retrieve each case and to be used as dependencies for the endpoints, respectively.

    We can define a separate module, called the `rbac.py` module. Let's start with the imports:

```
from typing import Annotated
from fastapi import (
    APIRouter,
    Depends,
    HTTPException,
    Status
)
```

```
from sqlalchemy.orm import Session

from db_connection import get_session
from models import Role
from security import (
    decode_access_token,
    oauth2_scheme
)
```

Then, we create the request model that we will use with the endpoints:

```
class UserCreateResquestWithRole(BaseModel):
    username: str
    email: EmailStr
    role: Role
```

Then, we define a support function to retrieve the user based on the token:

```
def get_current_user(
    token: str = Depends(oauth2_scheme),
    session: Session = Depends(get_session),
) -> UserCreateRequestWithRole:
    user = decode_access_token(token, session)
    if not user:
        raise HTTPException(
            status_code=status.HTTP_401_UNAUTHORIZED,
            detail="User not authorized",
        )

    return UserCreateRequestWithRole(
        username=user.username,
        email=user.email,
        role=user.role,
    )
```

We can then leverage this function to shortlist premium users only:

```
def get_premium_user(
    current_user: Annotated[
        get_current_user, Depends()
    ]
):
    if current_user.role != Role.premium:
        raise HTTPException(
```

```
                    status_code=status.HTTP_401_UNAUTHORIZED,
                    detail="User not authorized",
            )
        return current_user
```

2. Now, we can use the functions to create the respective endpoints with the router in the same module. First, we define a welcome page for all the users:

```
router = APIRouter()


@router.get(
    "/welcome/all-users",
    responses=..., # document responses
)
def all_users_can_access(
    user: Annotated[get_current_user, Depends()]
):
    return {
        f"Hello {user.username}, "
        "welcome to your space"
    }
```

Then, we define the endpoint, allowing only premium users:

```
@router.get(
    "/welcome/premium-user",
    responses={
        status.HTTP_401_UNAUTHORIZED: {
            "description": "User not authorized"
        }
    },
)
def only_premium_users_can_access(
    user: UserCreateResponseWithRole = Depends(
        get_premium_user
    ),
):
    return {
        f"Hello {user.username}, "
        "Welcome to your premium space"
    }
```

3. Let's add the router we create in `main.py`:

```
import security
import premium_access
import rbac

# rest of the module

app.include_router(premium_access.router)
app.include_router(rbac.router)

# rest of the module
```

We have just implemented two endpoints with permissions based on the use role.

To test our endpoints, start the server from the command line:

```
$ uvicorn main:app
```

Then, from your browser, go to the Swagger page at `http://localhost:8000/docs`, and you can see the new endpoints just created.

A way to experiment is to create a basic and a premium user and use the corresponding endpoints. After you have made the users, you can try using the `GET welcome/all-users` and `GET /welcome/premium-user` endpoints with both roles and see that the response matches the role's expectations.

In this recipe, you just made simple endpoints that are available based on the user role. You can also play around with making more roles and endpoints.

There's more...

Another way to apply RBAC is to assign a scope to a token. This scope can be a string that represents certain permissions. As a result, the role is controlled by the token generation system. In FastAPI, you can define scopes within the token. You can check out the dedicated documentation page for more information: `https://fastapi.tiangolo.com/advanced/security/oauth2-scopes/`.

Using third-party authentication

Incorporating third-party authentication into your FastAPI application allows users to log in using their existing social media accounts, such as Google or Facebook. This recipe guides you through the process of integrating GitHub third-party login, enhancing user experience by simplifying the sign-in process.

Getting ready

We'll focus on integrating GitHub OAuth2 for authentication. GitHub provides comprehensive documentation and a well-supported client library that simplifies the integration process.

You will need the `httpx` package in your environment, so if you haven't installed it with the `requirements.txt`, you can do it by running the following:

```
$ pip install httpx
```

You will need also a GitHub account set up. If you don't have one, create one; you can find a comprehensive guide on the official documentation at `https://docs.github.com/en/get-started/start-your-journey/creating-an-account-on-github`.

Then, you need to create an application in your account by following the following steps:

1. From your personal page, click on the profile icon on the top right of the screen, the navigate to **Settings | Developer settings | OAuth Apps | New OAuth App** and fill the required fields in the form:

    - **Application name**: For example, `SaasFastAPIapp`.

    - **Homepage URL**: The address of your SaaS home page at `http://localhost:8000/home`, which we will create later.

    - **Authorization callback URL**: This is our application's endpoint, which will be called to refresh the token. You can set it to `http://localhost:8000/github/auth/token`, which we will define later as well.

2. Click on **Register application** and the app will be created, and you will be redirected to a page listing essential data about your OAuth2 app.

3. Take note of the client ID and click on the **Generate a new client secret**.

4. Store the client secret you just created. With the client ID and the client secret, we can proceed to implement the third-party authentication by GitHub.

Now, we have all we need to integrate the GitHub third-party login with our application.

How to do it...

Let's start by creating a new module called `third_party_login.py` to store helper data and functions for the GitHub authentication. Then let's continue as follows.

1. Within the `third_party_login.py` module, you can define the variables used for the authentication:

```
GITHUB_CLIENT_ID = "your_github_client_id"
GITHUB_CLIENT_SECRET = (
    "your_github_client_secret"
)
GITHUB_REDIRECT_URI = (
    "http://localhost:8000/github/auth/token"
)
GITHUB_AUTHORIZATION_URL = (
    "https://github.com/login/oauth/authorize"
)
```

For `GITHUB_CLIENT_ID` and `GITHUB_CLIENT_SECRET`, use the values of your OAuth app.

> **Warning**
>
> In a production environment, make sure to not hardcode any username or client ID in your code base.

2. Then, still in the `third_party_login.py` module, let's define a helper function, `resolve_github_token`, that resolves the GitHub token and returns information about the user:

```
import httpx
from fastapi import Depends, HTTPException
from fastapi.security import OAuth2
from sqlalchemy.orm import Session

from models import User, get_session
from operations import get_user


def resolve_github_token(
    access_token: str = Depends(OAuth2()),
    session: Session = Depends(get_session),
) -> User:
    user_response = httpx.get(
```

```
        "https://api.github.com/user",
        headers={"Authorization": access_token},
    ).json()
    username = user_response.get("login", " ")
    user = get_user(session, username)
    if not user:
        email = user_response.get("email", " ")
        user = get_user(session, email)
    # Process user_response
    # to log the user in or create a new account
    if not user:
        raise HTTPException(
            status_code=403, detail="Token not valid"
        )
    return user
```

3. In a new module called `github_login.py`, we can start creating the endpoints used for the GitHub authentication. Let's create a new router and the `github_login` endpoint that will return the URL used by the frontend to redirect the user to the GitHub login page:

```python
import httpx
from fastapi import APIRouter, HTTPException, status

from security import Token
from third_party_login import (
    GITHUB_AUTHORIZATION_URL,
    GITHUB_CLIENT_ID,
    GITHUB_CLIENT_SECRET,
    GITHUB_REDIRECT_URI,
)

router = APIRouter()


@router.get("/auth/url")
def github_login():
    return {
        "auth_url": GITHUB_AUTHORIZATION_URL
        + f"?client_id={GITHUB_CLIENT_ID}"
    }
```

4. Now, let's add the router to the server in the `main.py` module:

    ```
    import github_login

    # rest of the module

    app.include_router(github_login.router)

    # rest of the module
    ```

5. Spin up the server with the same command, `uvicorn main:app`, and call the endpoint
    GET `/auth/url` we have just created. You will have a similar link in the response: `https://`
    `github.com/login/oauth/authorize?client_id=your_github_client_id`.

    This link is used by GitHub for the authentication. The redirection is managed by the frontend
    and is out of the scope of this book.

6. After validating the login, you will be redirected to a 404 page. This is because we still haven't
    created the callback endpoint in our application. Let's do so in the `github_login.py` module:

    ```
    @router.get(
        "/github/auth/token",
        response_model=Token,
        responses=..., # add responses documentation
    )
    async def github_callback(code: str):
        token_response = httpx.post(
            "https://github.com/login/oauth/access_token",
            data={
                "client_id": GITHUB_CLIENT_ID,
                "client_secret": GITHUB_CLIENT_SECRET,
                "code": code,
                "redirect_uri": GITHUB_REDIRECT_URI,
            },
            headers={"Accept": "application/json"},
        ).json()
        access_token = token_response.get("access_token")
        if not access_token:
            raise HTTPException(
                status_code=401,
                detail="User not registered",
            )
        token_type = token_response.get(
            "token_type", "bearer"
    ```

```
    )

    return {
        "access_token": access_token,
        "token_type": token_type,
    }
```

The endpoint we just created returns the actual access token.

7. If you restart the server and try to validate the GitHub login again with the link provded by the GET /auth/url endpoint, you will receive a response containing the token similar to the following:

```
{
    "access_token": "gho_EnHbcmHdCHD1Bf2QzJ2B6gyt",
    "token_type": "bearer"
}
```

8. The last piece of the puzzle is to create the home page endpoint that can be accessed with the GitHub token and will recognize the user by resolving the token. We can define it in the main.py module:

```
from third_party_login import resolve_github_token

@router.get(
    "/home",
    responses=…, # add responses documentation
)
def homepage(
    user: UserCreateResponse = Depends(
        resolve_github_token
    ),
):
    return {
        "message" : f"logged in {user.username} !"
    }
```

You've just implemented an endpoint that authenticates with the GitHub third-party authenticator.

How it works...

First, by using the register endpoint POST /register/user, add a user that has the same username or email as the GitHub account you are going to test.

Then, retrieve the token from the GitHub URL provided by the `GET /auth/url` endpoint.

You will use a token with your favorite tool to query the `GET /home` endpoint, which uses the GitHub token to validate permission.

At the time of writing, we cannot test endpoints requiring external bearer tokens with the interactive documentation, so feel free to use your favorite tool to query the endpoint by providing the bearer token in the headers authorization.

You can do it with **Postman**, for example, and you can use the equivalent `curl` request from your shell as well, as follows:

```
$ curl --location 'http://localhost:8000/home' \
--header 'Authorization: Bearer <github-token>'
```

If everything is correctly set up, you will receive the response:

```
{"message":"logged in <your-username> !"}
```

You just implemented and tested authentication by using a third-party application such as GitHub. Other providers such as Google or Twitter follow similar procedures, with small differences. Feel free to implement them as well.

See also

Take a look at the GitHub documentation that provides a guide on how to set up OAuth2 authentication:

- *GitHub OAuth2 integration*: `https://docs.github.com/en/apps/oauth-apps/building-oauth-apps/authorizing-oauth-apps`

You can use third-party authorization login with other providers that allow a similar configuration. You can check, for example, Google and Twitter:

- *Google OAuth2 integration*: `https://developers.google.com/identity/protocols/oauth2`

- *Twitter OAuth2 integration*: `https://developer.twitter.com/en/docs/authentication/oauth-2-0`

Implementing MFA

MFA adds a layer of security by requiring users to provide two or more verification factors to gain access to a resource. The recipe guides you through adding MFA to your FastAPI application, enhancing security by combining something the user knows (their password) with something they have (a device).

Getting ready

For our FastAPI application, we'll use a **time-based one-time password** (**TOTP**) as our MFA method. TOTP provides a six to eight-digit number that's valid for a short period, typically 30 seconds.

First, ensure you have the necessary packages installed:

```
$ pip install pyotp
```

Pyotp is a Python library that implements one-time password algorithms, including TOTP.

To use the TOTP authentication, we need to modify the user table in our database to take into account the TOTP secret used to validate the secret number.

Let's modify the User class in the models.py module by adding the totp_secret field:

```
class User(Base):
    # existing fields

    totp_secret: Mapped[str] = mapped_column(
        nullable=True
    )
```

We are now ready to implement MFA.

How to do it...

Let's start by creating two helper functions to generate a TOTP secret and TOTP URI used by the authenticatorthrough the following steps.

1. We define the functions in a new module called mfa.py:

    ```
    import pyotp


    def generate_totp_secret():
        return pyotp.random_base32()


    def generate_totp_uri(secret, user_email):
        return pyotp.totp.TOTP(secret).provisioning_uri(
            name=user_email, issuer_name="YourAppName"
        )
    ```

The TOTP URI can be a QR code as well in the form of a link.

We will use the `generate_totp_secret` and `generate_totp_uri` functions to create the endpoint to request MFA.

2. The endpoint will return a **TOTP URI** for use by the authenticator. To show the mechanism, we will also return the secret number, which in a real-life scenario is the number generated by the authenticator:

```python
from fastapi import (
    APIRouter,
    Depends,
    HTTPException,
    status,
)
from sqlalchemy.orm import Session

from db_connection import get_session
from operations import get_user
from rbac import get_current_user
from responses import UserCreateResponse


router = APIRouter()


@router.post("/user/enable-mfa")
def enable_mfa(
    user: UserCreateResponse = Depends(
        get_current_user
    ),
    db_session: Session = Depends(get_session),
):
    secret = generate_totp_secret()
    db_user = get_user(db_session, user.username)
    db_user.totp_secret = secret
    db_session.add(db_user)
    db_session.commit()
    totp_uri = generate_totp_uri(secret, user.email)

    # Return the TOTP URI
    # for QR code generation in the frontend
    return {
```

```
        "totp_uri": totp_uri,
        "secret_numbers": pyotp.TOTP(secret).now(),
    }
```

3. Now, we can create the endpoint to validate the secret number:

```
@app.post("/verify-totp")
def verify_totp(
    code: str,
    username: str,
    session: Session = Depends(get_session),
):
    user = get_user(session, username)
    if not user.totp_secret:
        raise HTTPException(
            status_code=status.HTTP_400_BAD_REQUEST,
            detail="MFA not activated",
        )

    totp = pyotp.TOTP(user.totp_secret)
    if not totp.verify(code):
        raise HTTPException(
            status_code=status.HTTP_401_UNAUTHORIZED,
            detail="Invalid TOTP token",
        )
    # Proceed with granting access
    # or performing the sensitive operation
    return {
        "message": "TOTP token verified successfully"
    }
```

As before, you need to include the router in the `FastAPI` object class in `main.py` for all the previous endpoints.

To test it, as usual spin up the server from the terminal by running:

```
$ uvicorn main:app
```

Make sure you have a user in your database, go to the interactive documentation, and call the /user/ enable-mfa endpoint by authenticating with the user credentials. You will get a response with the TOTP URI and a temporary secret number, like the following:

```
{
  "totp_uri":
  "otpauth://totp/YourAppName:giunio%40example.com?secret=
  NBSUC4CFDUT5IEYX4IR7WKBTDTU7LN25&issuer=YourAppName",
  "secret_numbers": "853567"
}
```

Take note of the secret number to use as a parameter of the /verify-totp endpoint with the username, and you will get this response:

```
{
  "message": "TOTP token verified successfully"
}
```

You've just implemented MFA in your FastAPI application and enhanced security by ensuring that even if a user's password is compromised, an attacker still needs access to the user's second factor (the device running the MFA app) to gain access.

See also

Take a look at the Python One-Time Password library in the official documentation:

- *Python One-Time Password library*: https://pyauth.github.io/pyotp/

Handling API key authentication

API key authentication is a simple yet effective way to control access to an application. This method involves generating a unique key for each user or service that needs access to your API and requiring that key to be included in the request headers.

API keys can be generated in various ways, depending on the level of security needed.

FastAPI doesn't have built-in support for API key authentication, but you can easily implement it using dependencies or middleware. A dependency is more flexible for most use cases, so we'll use that approach.

This recipe will show you a basic, yet not secure, way to implement it.

Getting ready

We will keep working on our application. However, you can apply this recipe to a simple application from scratch as well.

How to do it...

Let's create an `api_key.py` module to store the logic to handle API keys. The package will contain the API list and verification method:

```
from fastapi import HTTPException
from typing import Optional

VALID_API_KEYS = [
    "verysecureapikey",
    "anothersecureapi",
    "onemoresecureapi",
]


async def get_api_key(
    api_key: Optional[str]
):
    if (
        api_key not in VALID_API_KEYS
    ):
        raise HTTPException(
            status_code=403, detail="Invalid API Key"
        )
    return api_key
```

In the example, the keys are hardcoded into the `VALID_API_KEYS` list. However, in real-life production scenarios, the management and validation of the keys are usually done by dedicated libraries or even services.

Let's create an endpoint that makes use of the API key:

```
from fastatpi import APIrouter

router = APIRouter()

@router.get("/secure-data")
async def get_secure_data(
```

```
    api_key: str = Depends(get_api_key),
):
    return {"message": "Access to secure data granted"}
```

Now, add the router to the `FastAPI` object class in `main.py`, and then the endpoint is ready to be tested.

Spin up the server from the command by running the following:

```
$ uvicorn main:app
```

Go to the interactive documentation at `http://localhost:8000/docs` and test the endpoint you just created by providing an API key.

As you see, by adding a simple dependency to the endpoint, you can secure any endpoint of your app with an API key.

There's more...

We have developed a simple module for managing the API of our application. In production environment this can be handled by external services often provided by the hosting platform. However, If you are going to implement your API management system, keep in mind the best practices for API key authentication:

- **Transmission security**: Always use HTTPS to prevent API keys from being intercepted during transmission

- **Key rotation**: Regularly rotate API keys to minimize the risk of compromise

- **Limit permissions**: Assign minimal permissions required for each API key based on the principle of least privilege

- **Monitoring and revocation**: Monitor the usage of API keys and have mechanisms in place to revoke them if suspicious activity is detected

Handling session cookies and logout functionality

Managing user sessions and implementing logout functionality is crucial for maintaining security and user experience in web applications. This recipe shows how to handle session cookies in FastAPI, from creating cookies upon user login to securely terminating sessions upon logout.

Getting ready

Sessions provide a way to persist user data across requests. When a user logs in, the application creates a session on the server side and sends a session identifier to the client, usually in a **cookie**. The client sends this identifier back with each request, allowing the server to retrieve the user's session data.

The recipe will show how to manage cookies for sessions with login and logout functionality.

How to do it...

Cookies in FastAPI are easily managed by the Request and Response object classes. Let's create a login and a logout endpoints to attaches a session cookie to the response and ignore it from the request.

Let's create a dedicated module called user_session.py and add the /login endpoint:

```python
from fastapi import APIRouter, Depends, Response
from sqlalchemy.orm import Session

from db_connection import get_session
from operations import get_user
from rbac import get_current_user
from responses import UserCreateResponse

router = APIRouter()


@router.post("/login")
async def login(
    response: Response,
    user: UserCreateResponse = Depends(
        get_current_user
    ),
    session: Session = Depends(get_session),
):
    user = get_user(session, user.username)

    response.set_cookie(
        key="fakesession", value=f"{user.id}"
    )
    return {"message": "User logged in successfully"}
```

Testing the login endpoint won't be possible with the Swagger documentation because we need to verify that the fakesession cookie has been created.

Spin up the server with `uvicorn main:app` and use Postman to create a `Post` request to the `/login` endpoint by providing the authentication token for the user you want to log in.

Verify that the response contains the `fakesession` cookie by selecting **Cookies** from the drop-down menu of the response section.

Accordingly, we can define a logout endpoint that won't return any session cookie in the response:

```
@router.post("/logout")
async def logout(
    response: Response,
    user: UserCreateResponse = Depends(
        get_current_user
    ),
):
    response.delete_cookie(
        "fakesession"
    )  # Clear session data
    return {"message": "User logged out successfully"}
```

That's all you need to manage sessions.

To test the `POST /logout` endpoint, restart the server with `uvicorn`. Then, when calling the endpoint, make sure you provide the `fakesession` cookie in your HTTP request with the user bearer token. If you previously called the login endpoint, it should be automatically stored; otherwise, you can set it in the `Cookies` section of the request.

Check the response and confirm that the `fakesession` cookie is not present anymore in the response.

There's more...

There is a lot to learn about cookies besides the basic recipe. In a real-world setting, you can use specialized libraries or even external services.

Whatever your choice is, put security first and follow those practices to make your session secure and efficient:

- **Secure cookies**: Mark session cookies as `Secure`, `HttpOnly`, and `SameSite` to protect against **cross-site request forgery** (**CSRF**) and **cross-site scripting** (**XSS**) attacks

- **Session expiry**: Implement session expiry both in your session store and by setting a maximum age on the cookie

- **Regenerate Session ID**: Regenerate session IDs upon login to prevent session fixation attacks

- **Monitor sessions**: Implement mechanisms to monitor active sessions and detect anomalies

By integrating session management and logout functionality into your FastAPI application, you ensure that user state is managed securely and efficiently across requests. This enhances both the security and user experience of your application. Remember to follow best practices for session security to protect your users and their data effectively.

In the next chapter, we will see how to efficiently debug your FastAPI application.

See also

You can see more on managing cookies in Fast on the documentation page:

- *Response cookies*: `https://fastapi.tiangolo.com/advanced/response-cookies/`

5

Testing and Debugging FastAPI Applications

In this chapter of our journey through mastering FastAPI, we pivot towards a crucial aspect of software development that ensures the reliability, robustness, and quality of your applications: testing and debugging. As we delve into this chapter, you'll be equipped with the knowledge and tools necessary to create an effective testing environment, write and execute comprehensive tests, and debug your FastAPI applications with efficiency and precision.

Understanding how to properly test and debug is not just about finding errors; it's about ensuring your application can withstand real-world use, manage high traffic without faltering, and provide a seamless user experience. By mastering these skills, you'll be able to confidently enhance your applications, knowing that each line of code has been scrutinized and each potential bottleneck has been addressed.

We are going to create a proto application with a minimal setup to test the recipes.

By the end of this chapter, you will not only have a deep understanding of the testing frameworks and debugging strategies suitable for FastAPI but also practical experience in implementing these techniques to build more resilient applications. This knowledge is invaluable, as it directly impacts the quality of your software, its maintenance, and its scalability.

In this chapter we're going to cover the following recipes:

- Setting up testing environments
- Writing and running unit tests
- Testing API endpoints
- Handling logging messages
- Debugging techniques
- Performance testing for high traffic application

Technical requirements

To dive into the chapter and follow along with the recipes, ensure your setup includes the following essentials:

- **Python**: Make sure to have a Python version 3.7 or higher installed on your computer.

- **FastAPI**: Have `fastapi` package in your working environment.

- **Pytest**: Be familiar with `pytest` framework, which is a testing framework largely used to test Python code.

The code used in the chapter is hosted on GitHub at the address: `https://github.com/PacktPublishing/FastAPI-Cookbook/tree/main/Chapter05`.

You can setup a virtual environment for the project within the project root folder is also recommended to manage dependencies efficiently and maintain project isolation. Within your virtual environment, you can install all the dependencies at once by using the `requirements.txt` provided on the GitHub repository in the project folder:

```
$ pip install -r requirements.txt
```

A basic knowledge of HTTP protocol, although not required, can be beneficial.

Setting up testing environments

This recipe will show you how to setup an efficient and effective testing environment tailored for FastAPI applications. By the end of the recipe, you will have a solid foundation for writing, running, and managing tests.

Getting ready

Make sure you have an application running. If not you can start by creating a project folder `proto_app`.

If you haven't installed the packages with the requirements.txt file provided on the GitHub repository, then install the testing libraries `pytest` and `httpx` in your environment with:

```
$ pip install pytest pytest-asyncio httpx
```

In the project root folder create a new folder `proto_app` with a `main.py` module containing the app object instance:

```
from fastapi import FastAPI

app = FastAPI()
```

```
@app.get("/home")
async def read_main():
    return {"message": "Hello World"}
```

With a minimal app setup, we can proceed by scaffolding our project to accommodate the tests.

How to do it...

First, let's start by structuring our project folder tree to accommodate tests.

1. In the root directory let's create a `pytest.ini` file and a `tests` folder containing the test module `test_main.py`. The project structure should look like this:

   ```
   protoapp/
   |– protoapp/
   |   |– main.py
   |– tests/
   |   |– test_main.py
   |– pytest.ini
   ```

2. The `pytest.ini` contains instructions for `pytest`. You can write in it:

   ```
   [pytest]
   pythonpath = . protoapp
   ```

   This will add the project root and the folder `protoapp`, containing the code, to the PYTHONPATH when running `pytest`.

3. Now, in the `test_main.py` module, let's write a test for the `/home` endpoint we created earlier:

   ```python
   import pytest
   from httpx import ASGITransport, AsyncClient

   from protoapp.main import app

   @pytest.mark.asyncio
   async def test_read_main():
       client = AsyncClient(
           transport=ASGITransport(app=app),
           base_url="http://test",
       )
       response = await client.get("/home")
       assert response.status_code == 200
       assert response.json() == {
   ```

```
        "message": "Hello World"
    }
```

As a first check of the environment, we can try to collect the tests. From the `protoapp` root project folder run:

```
$ pytest --collect-only
```

You should get an output like:

```
configfile: pytest.ini
plugins: anyio-4.2.0, asyncio-0.23.5, cov-4.1.0
asyncio: mode=Mode.STRICT
collected 1 item


<Dir protoapp>
  <Dir tests>
    <Module test_main.py>
      <Coroutine test_read_main>
```

This specifies:

* The configuration file `pytest.ini`

* The `pytest` plugins used

* The directory tests, the module `test_main.py` and the test `test_read_main` which is a coroutine

4. Now, from the command line terminal at the project root folder level, run the `pytest` command:

```
$ pytest
```

You've just setup the environment to test our proto application.

See also

The recipe has shown how to configure `pytest` within a **FastAPI** project with some of the good practices. Feel free to dig deeper into the **Pytest** official documentation at the links:

* *Pytest configuration*: `https://docs.pytest.org/en/stable/reference/customize.html`

* *Setup PYTHONPATH in Pytest*: `https://docs.pytest.org/en/7.1.x/explanation/pythonpath.html`

* *Pytest good practices*: `https://docs.pytest.org/en/7.1.x/explanation/goodpractices.html`

Writing and running unit tests

Once we setup our testing environment, we can focus on the process of writing and executing tests for FastAPI applications. Unit tests are essential for validating the behaviour of individual parts of your application in isolation, ensuring they perform as expected. In this recipe, you will learn to test the endpoints of your application.

Getting ready

We will use `pytest` to test the FastAPI client in unit tests. Since the recipe will utilize common testing *fixtures*, used in most **Python** standard code, make sure to be familiar with the test fixtures before diving into the recipe. If this is not the case, you can always refer to the dedicated documentation page at the link: `https://docs.pytest.org/en/7.1.x/reference/fixtures.html`.

How to do it...

We will start by creating a unit test for the same `GET /home` endpoint, but differently from the previous recipe. We will use the `TestClient` class provided by FastAPI.

Let's create a fixture for that. Since it could be used by multiple tests let's do it in a new `conftest.py` module under the `tests` folder. The `conftest.py` is a default file used by `pytest` to store common elements shared amongst test modules.

In the `conftest.py` let's write:

```python
import pytest
from fastapi.testclient import TestClient

from protoapp.main import app

@pytest.fixture(scope="function")
def test_client(db_session_test):
    client = TestClient(app)
    yield client
```

We are now ready to leverage the `test_client` fixture to create a proper unit test for our endpoint.

We will write our test in the `test_main.py` module:

```python
def test_read_main_client(test_client):
    response = test_client.get("/home")
    assert response.status_code == 200
    assert response.json() == {"message": "Hello World"}
```

And that's it. Compared to the previous test, this one is more compact and faster to write, thanks to the `TestClient` class provided by FastAPI package.

Now run `pytest`:

```
$ pytest
```

You will see a message on the terminal showing that two tests have been collected and run successfully.

See also

You can check more on the test client for FastAPI in the official documentation:

- *FastAPI Test Client*: `https://fastapi.tiangolo.com/reference/testclient/`

Testing API Endpoints

Integration tests verify that different parts of your application work together as expected. They are crucial for ensuring that your system's components interact correctly, especially when dealing with external services, databases, or other APIs.

In this recipe, we will test two endpoints that interact with an SQL database. One will add an item to the database, the other will read an item based on the ID.

Getting ready

To apply the recipe you need your testing environment already setup for `pytest`. If this is not the case check the recipe *Setting up testing environments* of the same chapter.

Also, the recipe will show you how to make integration tests with existing endpoints of the application. You can use it for your application or you can build the endpoints for our `protoapp` as follows.

If you are using the recipe to test your endpoint you can directly jump on the *How to it…* section and apply the rules to tour endpoints.

Otherwise, If you haven't installed the packages from the `requirements.txt`, install `sqlalchemy` package in your environment:

```
$ pip install "sqlalchemy>=2.0.0"
```

Now let's setup the database connection through the following steps.

1. Under the protoapp folder, at the same level as the main.py module, let's create a module database.py containing the setup of the database. Let's start by creating the Base class:

    ```
    from sqlalchemy.orm import DeclarativeBase,

    class Base(DeclarativeBase):
        pass
    ```

    We will use the Base class to define the Item mapping class.

2. Then the database Item mapping class will be like:

    ```
    from sqlalchemy.orm import (
        Mapped,
        mapped_column,
    )

    class Item(Base):
        __tablename__ = "items"
        id: Mapped[int] = mapped_column(
            primary_key=True, index=True
        )
        name: Mapped[str] = mapped_column(index=True)
        color: Mapped[str]
    ```

3. Then, we define the database engine that will handle the session:

    ```
    DATABASE_URL = "sqlite:///./production.db"

    engine = create_engine(DATABASE_URL)
    ```

    The engine object will be used to handle the session.

4. Then, let's bind the engine to the Base mapping class:

    ```
    Base.metadata.create_all(bind=engine)
    ```

    Now the engine can map the database table to our Python classes.

5. Last in the `database.py` module let's create a `SessionLocal` class that will generate the session as:

```
SessionLocal = sessionmaker(
    autocommit=False, autoflush=False, bind=engine
)
```

The `SessionLocal` is a class that will initialize the database session object.

6. Finally, before creating the endpoints, let's create a database session.

Since the app is relatively small, we can do it the same `main.py`:

```
from protoapp.database import SessionLocal


def get_db_session():
    db = SessionLocal()
    try:
        yield db
    finally:
        db.close()
```

We will use the session to interact with the database.

Now that we have setup the database connection, in the `main.py` module, we can create the endpoints one to add an item to the database and one to read it. Let's do it as follows.

1. Let's start by creating the request body for the endpoints as::

```
from pydantic import BaseModel


class ItemSchema(BaseModel):
    name: str
    color: str
```

2. The endpoint used to add an item will then be:

```
from fastapi import (
    Depends,
    Request,
    HTTPException,
    status
)
from sqlalchemy.orm import Session
```

```
@app.post(
"/item",
response_model=int,
status_code=status.HTTP_201_CREATED
)
def add_item(
    item: ItemSchema,
    db_session: Session = Depends(get_db_session),
):
    db_item = Item(name=item.name, color=item.color)
    db_session.add(db_item)
    db_session.commit()
    db_session.refresh(db_item)
    return db_item.id
```

The endpoint will return the item ID affected when the item is stored in the database.

3. Now that we have the endpoint to add the item, we can proceed by creating the endpoint to retrieve the item based on its ID:

```
@app.get("/item/{item_id}", response_model=ItemSchema)
def get_item(
    item_id: int,
    db_session: Session = Depends(get_db_session),
):
    item_db = (
        db_session.query(Item)
        .filter(Item.id == item_id)
        .first()
    )
    if item_db is None:
        raise HTTPException(
            status_code=404, detail="Item not found"
        )

    return item_db
```

If the ID does not correspond to any item in the database the endpoint will return a 404 status code.

We have just created the endpoints that will allow us to create an integration test.

How to do it...

Once we have the endpoints, in the `tests` folder we should adapt our `test_client` fixture to use a different session than the one used in production.

We will break the process into two main actions:

- Adapt the test client to accommodate the testing database session
- Create the test to simulate the interaction of the endpoints

Let's do it by following these steps.

1. First, In the `conftest.py` file we created earlier in the recipe *Writing and running unit tests*, let's define a new engine that will use an in-memory SQLite database and bind it to the mapping `Base` class:

```
from sqlalchemy.pool import StaticPool
from sqlalchemy import create_engine

engine = create_engine(
    "sqlite:///:memory:",
    connect_args={"check_same_thread": False},
    poolclass=StaticPool,
)


Base.metadata.create_all(bind=engine)    # Bind the engine
```

2. Let's create a dedicated session maker for the testing session as:

```
from sqlalchemy.orm import sessionmaker

TestingSessionLocal = sessionmaker(
    autocommit=False, autoflush=False, bind=engine
)
```

3. Similarly to the function `get_db_session` in the `main.py` module, we can create a fixture to retrieve the test session in the conftest.py module:

```
@pytest.fixture
def test_db_session():
    db = TestingSessionLocal()
    try:
        yield db
```

```
    finally:
        db.close()
```

4. Then, we should modify the `test_client` to use this session instead of the production one. We can do it by overwriting the dependency that returns the session with the one we just created. FastAPI allows you to do it easily by calling the test client's method `dependency_overrides` as:

```
from protoapp.main import app, get_db_session

@pytest.fixture(scope="function")
def test_client(test_db_session):
    client = TestClient(app)
    app.dependency_overrides[get_db_session] = (
        lambda: test_db_session
    )

    return client
```

Each time the test client needs to call the session, the fixture will replace it with the test session that uses the in-memory database.

5. Then, to verify the interaction of our application with the database, we create a test that:

*   Create the item into the database through the POST `/item` endpoint

*   Verify that the item is correctly created into the test database by using the test session

*   Retrieve the item through the GET `/item` endpoint

You can put the test into the `test_main.py` and here is how it would look like:

```
def test_client_can_add_read_the_item_from_database(
    test_client, test_db_session
):
    response = test_client.get("/item/1")
    assert response.status_code == 404

    response = test_client.post(
        "/item", json={"name": "ball", "color": "red"}
    )
    assert response.status_code == 201
    # verify the user was added to the database
    item_id = response.json()
    item = (
        test_db_session.query(Item)
```

```
        .filter(Item.id == item_id)
        .first()
    )
    assert item is not None

    response = test_client.get(f"item/{item_id}")
    assert response.status_code == 200
    assert response.json() == {
        "name": "ball",
        "color": "red",
    }
```

You've just created an integration test for our proto application, feel free to enrich your application and create more tests accordingly.

See also

We have setup an in-memory SQLite database for our tests. Since each session is bonded to a thread, the engine needs to be configured accordingly to not flush data.

The configuration strategy has been found on the following documentation page:

- *SQLite In-Memory Database Configuration*: https://docs.sqlalchemy.org/en/14/dialects/sqlite.html#using-a-memory-database-in-multiple-threads

Running tests techniques

By systematically covering all endpoints and scenarios, you ensure that your API behaves correctly under various conditions, providing confidence in your application's functionality. Thoroughly testing API endpoints is essential for building reliable and robust applications.

The recipe will explain to you how to run tests individually or by group and how to check the test coverage of our code.

Getting ready

To run the recipe, make sure you already have some tests in place, or you already followed all the previous recipes of the chapter. Also, make sure you have your PYTHONPATH for tests defined in your pytest.ini. Have a look at the recipe *Setting up testing environments* on how to do it.

How to do it...

We will start by looking at how to run tests by default grouping (individually or by module), and then we will cover a technique for customizing test grouping based on marks.

As you already know, all unit tests can be run from the terminal with the command:

```
$ pytest
```

However, a test can be run individually according to the test call syntax:

```
$ pytest <test_module>.py::<test_name>
```

For example, if we want to run the test function `test_read_main_client`, run:

```
$ pytest tests/test_main.py::test_read_main
```

Sometimes test names become too complicated to remember or we have a specific need to run only a targeted set of tests. Here is where test marks come to the aid.

Let's imagine we want to run only integration tests. In our app, the only integration test is represented by the function `tests_client_can_add_read_the_item_from_database`.

We can apply a mark by adding the specific decorator to the function:

```
@pytest.mark.integration
def test_client_can_add_read_the_item_from_database(
    test_client, test_db_session
):
    # test content
```

Then, in the `pytest.ini` configuration add the `integration` marker in the dedicated sections to register the mark:

```
[pytest]
pythonpath = protoapp .
markers =
    integration: marks tests as integration
```

Now you can run the targeted tests by running:

```
$ pytest -m integration -vv
```

In the output message, you will see that only the marked test has been selected and run. You can use markers to group your application's tests based on logical criteria, for example by functional meaning one group for **create, read, update and delete (CRUD)** operations, one group for security operations, and so on.

Check test coverage

To make sure that your endpoints are covered by testing as well as the text lines of your code, it can become useful to have an idea of the test coverage.

Test coverage is a metric used in software testing to measure the extent to which the source code of a program is executed when a particular test suite runs.

To use it with `pytest`, if you didn't install the packages with the `requirements.txt`, you need to install `pytest--cov` package:

```
$ pip install pytest-cov
```

The way it works is quite straightforward. You need to pass the source code root, in our case the `protoapp` directory, to the parameter `--cov` of `pytest` and tests root folder, in our case tests as follows:

```
$ pytest --cov protoapp tests
```

You will see a table in the output listing the coverage percentage for each module:

```
Name                   Stmts   Miss   Cover
-----------------------------------------------
protoapp\database.py      16      0    100%
protoapp\main.py          29      4     86%
-----------------------------------------------
TOTAL                     45      4     91%
```

In addition, a file named `.coverage` has been created. This is a binary file containing data on the test coverage and that can be used with additional tools to generate reports out of it.

For example, if you run:

```
$ coverage html
```

It will create a folder `htmlcov` with an `index.html` page containing the coverage page and you can visualize it by opening it with a browser.

See also

You can check more on various options to invoke unit tests with Pytest and how to evaluate test coverage at the official documentation links

- *Invoke Unit test with Pytest*: https://docs.pytest.org/en/7.1.x/how-to/usage.html

- *Pytest Coverage*: https://pytest-cov.readthedocs.io/en/latest/

Handling logging messages

Effectively managing logs in application development not only aids in identifying errors promptly but also provides valuable insights into user interactions, system performance, and potential security threats. It serves as a crucial tool for auditing, compliance, and optimizing resource utilization, ultimately enhancing the reliability and scalability of the software.

This recipe will show how to efficiently implement a logging system into our FastAPI application to monitor the calls to the API.

Getting ready

We are going to use some basic features of the Python logging ecosystem.

Although the example is basic, you can refer to the official documentation to get familiar with related terms such as **logger, handler** , **formatter**, and **log level**. Follow this link:

`https://docs.python.org/3/howto/logging-cookbook.html`.

To implement logging into FastAPI, make sure you have a running application or use the `protoapp` we developed all along the chapter.

How to do it...

We want to create a logger that prints the client's calls information to the terminal and logs them into a file.

Let's create the logger into a dedicated `logging.py` module under the folder `protoapp`, through the following steps.

1. Let's start by defining the logger with a level value to `INFO`:

    ```
    import logging

    client_logger = logging.getLogger("client.logger")
    logger.setLevel(logging.INFO)
    ```

    Since we want to stream the message to the console and store it in a file, we will need to define two separate handlers.

2. Now let's define the handler to print log messages to the console. We will use a `StreamHandler` object from the `logging` built-in package:

    ```
    console_handler = logging.StreamHandler()
    ```

This will stream the message to the console.

3. Let's create a colorized formatter and add it to the handler we just created:

```
from uvicorn.logging import ColourizedFormatter

console_formatter = ColourizedFormatter(
    "%(levelprefix)s CLIENT CALL - %(message)s",
    use_colors=True,
)
console_handler.setFormatter(console_formatter)
```

The formatter will format log messages in the same of the default logger uvicorn logger used by FastAPI.

4. Then let's add the handler to the logger:

```
client_logger.addHandler(console_handler)
```

We have just set up the logger to print message to the console.

5. Let's repeat the previous *steps from 1 to 4* to create a handler that stores messages into a file and adds it to our `client_logger`:

```
from logging.handlers import TimedRotatingFileHandler

file_handler = TimedRotatingFileHandler("app.log")

file_formatter = logging.Formatter(
    "time %(asctime)s, %(levelname)s: %(message)s",
    datefmt="%Y-%m-%d %H:%M:%S",
)

file_handler.setFormatter(file_formatter)
client_logger.addHandler(file_handler)
```

Now we have our logger setup. Each message will be streamed to the console and stored in a `app.log` file.

6. Once we have built our `client_logger`, we need to use it in the code to get information about clients calls.

You can reach this by adding the logger and a dedicated middleware in the `main.py` module:

```python
from protoapp.logging import client_logger

# ... module content
@app.middleware("http")
async def log_requests(request: Request, call_next):
    client_logger.info(
        f"method: {request.method}, "
        f"call: {request.url.path}, "
        f"ip: {request.client.host}"
    )
    response = await call_next(request)
    return response
```

7. Now spin up the server:

```
$ uvicorn protoapp.main:app
```

Try to call any of the endpoints we defined, you will see on the terminal the logs we just defined for the request and response. Also, you will find only the messages from our `logger_client` in a newly created `app.log` file automatically created by the application.

There's more

Defining a proper logging strategy would require a separate cookbook and it is out of the scope of the book. However, when building a logger into an application it is important to follow some guidelines:

- **Use standard Logging Levels Appropriately**. A classical leveling system is made up of 4 levels: **INFO, WARNING, ERROR, CRITICAL**. You may need to have more or even less than four depending on the application. Anyway, place each message at the appropriate level.

- **Consist Log Format**. Maintain a consistent log format across your application. This includes consistent datetime formats, including the severity level, and describing the event clearly. A consistent format helps in parsing logs and automating log analysis.

- **Include Contextual Information**. Include relevant contextual information in your logs (e.g., user ID, transaction ID) to help trace and debug issues across your application's workflow.

- **Avoid Sensitive Information**. Never log sensitive information such as passwords, API keys, or **personal identifiable information (PII)**. If necessary, mask or hash these details.

- **Make Efficient Logging**. Be mindful of the performance impact of logging. Logging excessively can slow down your application and lead to log noise, making it hard to find useful information. Balance the need for information against the performance impact.

And of course, this is not a comprehensive list.

See also

Python distribution comes with a powerful built-in package for logging, feel to have a look at the official documentation:

- *Python logging*: `https://docs.python.org/3/library/logging.html`

Furthermore, discover more on logging best practices and guidelines at the **Sentry** blog:

- *Logging Guidelines*: `https://blog.sentry.io/logging-in-python-a-developers-guide/`

Sentry is a tool to monitor Python code.

Debugging techniques

Mastering debugging application development is crucial for identifying and fixing issues efficiently. This recipe delves into the practical use of the debugger, leveraging tools and strategies to pinpoint problems in your FastAPI code.

Getting ready

All you need to do to apply the recipe is to have a running application. We can keep on working with our `protoapp`.

How to do it...

The Python distribution already comes with a default debugger called pdb. If you use an **integrated development environment** (**IDE**), it usually comes with an editor distribution debugger. Whatever you are using to debug your code, you must be familiar with the concept of breakpoints.

A **breakpoint** is a point within the code that pauses the execution and shows you the state of the code variables and calls. It can be attached with a condition that, if satisfied, activate it or skips otherwise.

Whether you are using the Python distribution debugger pdb or the one provided by your IDE, it can be useful to define a starting script to spin up the server.

Create on the project root folder a file called `run_server.py` containing the following code:

```
import uvicorn
from protoapp.main import app


if __name__ == "__main__":
    uvicorn.run(app)
```

The script imports the `uvicorn` package and our application `app` and runs the application into the `uvicorn` server. It is equivalent to the launching command:

```
$ uvicorn protoapp.main:app
```

Having a script gives us more flexibility to run the server and include it into a broader python routine if required.

To check that it is correctly setup run the script as you would run a normal python script:

```
$ python run_server.py
```

With your favourite browser go to `localhost:8000/docs` and check that the documentation has been correctly generated.

Debugging with PDB

The PDB debugger comes by default with any Python distribution. From Python versions higher than 3.7, you can define a breakpoint by simply adding the function call `breakpoint()` at the line of the code you want to pause, and then run the code as you would it normally.

If you then run the code, when it reaches the breakpoint line, the execution will automatically shift to debug mode, and you can run debugging commands from the terminal. You can find the list of the commands you can run by typing help:

```
(Pdb) help
```

You can run commands to list variables, show the stack trace to check to recent frame, or define new breakpoints with conditions and more.

Here you can find the list of all the command available: `https://docs.python.org/3/library/pdb.html#debugger-commands`.

You can also invoke pdb as a module. In this case pdb will automatically enter **post-mortem** debugging if the program exists abnormally:

```
$ python -m pdb run_server.py
```

That means that pdb will restart the program automatically by preserving pdb module's execution state including breakpoints.

The same can be done when debugging tests by calling `pytest` as a module, for example:

```
$ python -m pdb -m pytest tests
```

Another debugging strategy consists of leveraging the reload functionality of the uvicorn server. To do that, you need to modify the run_server.py file as:

```
import uvicorn

if __name__ == "__main__":
    uvicorn.run("protoapp.main:app", reload=True)
```

Then, run the server without the pdb module:

```
$ python run_server.py
```

In this way, you can always use the breakpoints at ease with the reloading server functionality.

At the time of writing, **post-mortem** debugging is not supported with the automatic reload of unvicorn.

Debugging with VS Code

VS Code Python extension comes with its distribution debugger called *debugpy*. Configurations for the running environment can be managed in the .vscode/launch.json file. An example of the configuration file to debug our server is:

```
{
    "version": "0.2.0",
    "configurations": [
        {
            "name": "Python Debugger FastAPI server",
            "type": "debugpy",
            "request": "launch",
            "program": "run_server.py",
            "console": "integratedTerminal",
        },
    ]
}
```

The configuration specifies the type of debugger to use (debugpy), the program to run (our launching script run_server.py), and it can be found in the GUI options.

The request field specifies the mode to run the debugger, it can be *launch*, intended to run the program, or *attach*, intended to be attached to an already running instance, particularly useful to debug programs running on remote instances.

Debugging remote instance is out of the scope of the recipe, but you can find detailed instructions at on the official documentation: https://code.visualstudio.com/docs/python/debugging#_debugging-by-attaching-over-a-network-connection

Debugging configuration can be setup to run unit tests as well by leveraging the *Test Explorer* extension. The extension will look for a configuration in the `launch.json` containing `"type": "python"` and `"purpose": ["debug-test"]` (or `"request": "test"`). An example of configuration to debug tests would be:

```
{
    "version": "0.2.0",
    "configurations": [
        {
            "name": "Debug test",
            "type": "python",
            "request": "launch",
            "console": "integratedTerminal",
            "justMyCode": false,
            "stopOnEntry": true,
            "envFile": "${workspaceFolder}/.env.test",
            "purpose": ["debug-test"]
        }
    ]
}
```

You can find an extensive explication on the extension page from the VS Code marketplace at: `https://marketplace.visualstudio.com/items?itemName=LittleFoxTeam.vscode-python-test-adapter`.

Debugging with PyCharm

PyCharm manages code execution through run/debug configurations, which are sets of named startup properties detailing execution parameters and environments. These configurations allow running scripts with different settings, such as using various Python interpreters, environment variables, and input sources.

Run/debug configurations are of two kinds:

- Temporary: Automatically generated for each run or debug session.
- Permanent: Manually created from a template or by converting a temporary one, and saved within your project indefinitely until deleted.

PyCharm by default uses an existing permanent configuration or creates a temporary one for each session. Temporary configurations are capped at five, with the oldest deleted for new ones. This limit can be adjusted in the settings (**Settings** | **Advanced Settings** | **Run/Debug** | **Temporary configurations limit**). Icons distinguish between permanent (opaque) and temporary (semi-transparent) configurations.

Each configuration can be stored in a single xml file that is automatically detected by the GUI.

An example of configuration for our FastAPI `protoapp` is the following:

```
<component name="ProjectRunConfigurationManager">
  <configuration default="false" name="run_server"
    type="PythonConfigurationType" factoryName="Python"
    nameIsGenerated="true">
    <module name="protoapp" />
    <option name="INTERPRETER_OPTIONS" value="" />
    <option name="PARENT_ENVS" value="true" />
    <envs>
      <env name="PYTHONUNBUFFERED" value="1" />
    </envs>
    <option name="WORKING_DIRECTORY"
      value="$PROJECT_DIR$" />
    <option name="IS_MODULE_SDK" value="true" />
    <option name="ADD_CONTENT_ROOTS" value="true" />
    <option name="ADD_SOURCE_ROOTS" value="true" />
    <option name="SCRIPT_NAME"
      value="$PROJECT_DIR$/run_server.py" />
    <option name="SHOW_COMMAND_LINE" value="false" />
    <option name="MODULE_MODE" value="false" />
    <option name="REDIRECT_INPUT" value="false" />
    <option name="INPUT_FILE" value="" />
    <method v="2" />
  </configuration>
</component>
```

You can find a detailed guide on how to setup it at the dedicated Pycharm documentation page at: `https://www.jetbrains.com/help/pycharm/run-debug-configuration.html`.

See also

Feel free to dig into each of the debugging solutions and concepts we just explained at the links:

- *Python distribution debugger*: `https://docs.python.org/3/library/pdb.html`

- *Breakpoints*: `https://docs.python.org/3/library/functions.html#breakpoint`

- *Uvicorn Settings*: `https://www.uvicorn.org/settings/`

- *Debugging with VS Code*: `https://code.visualstudio.com/docs/python/debugging`

- *Debugy Debugger*: `https://github.com/microsoft/debugpy/`
- *Debugging with PyCharm*: `https://www.jetbrains.com/help/pycharm/debugging-your-first-python-application.html`

Performance testing for high traffic applications

Performance testing is crucial for ensuring your application can handle real-world usage scenarios, especially under high load. By systematically implementing and running performance tests, analyzing results, and optimizing based on findings, you can significantly improve your application's responsiveness, stability, and scalability.

The recipe will show the basics of how to benchmark your application with **Locust** framework.

Getting ready

To run performance testing you need a working application, we will use our `protoapp`, and a testing framework. We will use **Locust** framework for the purpose, which a testing framework based on Python syntax.

You can find a detailed explication on the official documentation at: `https://docs.locust.io/en/stable/`.

Before starting, make sure you installed it in your virtual environment by running:

```
$ pip install locust
```

Now we are ready to setup our configuration file and run the locust instance.

How to do it...

With the application running and the `locust` package installed, we will proceed by specifying our configuration to run the performance test.

Create a `locustfile.py` in your project root. This file will define the behavior of users interacting with your application under test.

A minimal example of `locustfile.py` can be:

```
from locust import HttpUser, task


class ProtoappUser(HttpUser):
    host = "http://localhost:8000"

    @task
```

```
def hello_world(self):
    self.client.get("/home")
```

The configuration defines a client class with the service address and the endpoint we want to test.

Start your FastAPI server with:

```
$ uvicorn protoapp.main:app
```

Then in another terminal window run locust:

```
$ locust
```

Open your browser and navigate to `http://localhost:8089` to access the web interface of the application.

The web interface is intuitively designed, making it straightforward to:

- **Set Concurrent Users**: Specify the maximum number of users accessing the service simultaneously during peak usage.
- **Configure Ramp-Up Rate**: Determine the rate of new users added per second to simulate increasing traffic.

After configuring these parameters, click the **Start** button to initiate a simulation that generates traffic to the protoapp via the `/home` endpoint defined in the `locustfile.py`.

Alternatively, you can simulate traffic using the command line. Here's how:

```
$ locust --headless --users 10 --spawn-rate 1
```

This command runs Locust in a headless mode to simulate:

- 10 users accessing your application concurrently.
- A spawn rate of 1 user per second.

You push your test experience further by including it in a **Continuous Integration /Continuous Delivery (CI/CD)** pipeline before deploying, or even into a larger testing routine.

Dig into the documentation to test every aspect of the traffic for your application.

You have all the tools to debug and fully test your application.

In the next chapter, we are going to build a comprehensive RESTful application interacting with an SQL database.

See also

You can find more on Locust on the official documentation pages:

- *Locust QuickStart*: `https://docs.locust.io/en/stable/quickstart.html`

- *Writing a Locust file*: `https://docs.locust.io/en/stable/writing-a-locustfile.html`

- *Running Locust from the Command Line*: `https://docs.locust.io/en/stable/running-without-web-ui.html`

6

Integrating FastAPI with SQL Databases

We'll now embark on a journey to harness the full potential of SQL databases within your FastAPI applications. This chapter is meticulously designed to guide you through the nuances of leveraging **SQLAlchemy**, a powerful SQL toolkit and **object-relational mapping** (**ORM**) for Python. From setting up your database environment to implementing sophisticated **create, read, update and delete** (**CRUD**) operations and managing complex relationships, this chapter provides a comprehensive blueprint for integrating SQL databases seamlessly with FastAPI.

By creating a basic ticketing platform, you'll practically engage in configuring SQLAlchemy with FastAPI, creating data models that reflect your application's data structures, and crafting efficient, secure CRUD operations.

Moreover, you'll explore the management of database migrations with **Alembic**, ensuring your database schema evolves alongside your application without hassle. This chapter doesn't stop at just handling data; it delves into optimizing SQL queries for performance, securing sensitive information within your database, and managing transactions and concurrency to ensure data integrity and reliability.

By the end of this chapter, you'll be adept at integrating and managing SQL databases in your FastAPI applications, equipped with the skills to ensure your applications are not only efficient and scalable but also secure. Whether you're building a new application from scratch or integrating a database into an existing project, the insights and techniques covered here will empower you to leverage the full power of SQL databases in your FastAPI projects.

In this chapter, we're going to cover the following recipes:

- Setting up SQLAlchemy
- Implementing CRUD operations
- Working with migrations
- Handling relationships in SQL databases

- Optimizing SQL queries for performance
- Securing sensitive data in SQL databases
- Handling transactions and concurrency

Technical requirements

To follow along with all the recipes of the chapter, make sure you have these essentials in your setup:

- **Python**: Your environment should have a Python version above 3.9 installed.
- **FastAPI**: It should be installed in your virtual environment with all the dependencies it needs. If you didn't do it in the previous chapters, you can easily do it from your terminal:

```
$ pip install fastapi[all]
```

The code that accompanies the chapter is available on GitHub at the following link: https://github.com/PacktPublishing/FastAPI-Cookbook/tree/main/Chapter06

It is also advisable to create a virtual environment for the project inside the project root folder, to handle dependencies well and keep the project separate. In your virtual environment, you can install all the dependencies at once by using the requirements.txt file from the GitHub repo in the project folder:

```
$ pip install -r requirements.txt
```

Since the code of the chapter will make use of the async/await syntax from the asyncio Python library, you should be already familiar with it. Feel free to read more about asyncio and async/await syntax at the following links:

- https://docs.python.org/3/library/asyncio.html
- https://fastapi.tiangolo.com/async/

Now that we have this ready Once we have everything ready, we can begin preparing our recipes.

Setting up SQLAlchemy

To begin any data application, you need to establish a database connection. This recipe will help you set up and configure sqlalchemy package with an **SQLite** database so that you can use the advantages of SQL databases in your applications.

Getting ready

The project is going to be fairly large, so we will put the working modules for the application in a folder named app, which will be under the root project folder that we will call ticketing_system.

You need `fastapi`, `sqlalchemy`, and `aiosqlite` installed in your environment to use the recipe. The recipe is meant to work with `sqlalchemy` with versions above 2.0.0. You can still use version 1; however, some adaptions are required. You can find a migration guide at the following link: `https://docs.sqlalchemy.org/en/20/changelog/migration_20.html`.

If you haven't installed the packages with the `requirements.txt` file in the repo, you can do it by running the following:

```
$ pip install fastapi[all] "sqlalchemy>=2.0.0" aiosqlite
```

Once the packages are correctly installed, you can follow the recipe.

How to do it...

The setup of a generic SQL database connection with `sqlalchemy` will go through the following steps:

1. Creating mapping object classes, that will match the database tables
2. Creating abstraction layers, an engine, and a session to communicate with the database
3. Initializing a database connection, at the server startup

Creating mapping object classes

In the app folder, let's create a module called `database.py` and then create a class object to track tickets as follows:

```python
from sqlalchemy import Column, Float, ForeignKey, Table
from sqlalchemy.orm import (
    DeclarativeBase,
    Mapped,
    mapped_column,
)


class Base(DeclarativeBase):
    pass


class Ticket(Base):
    __tablename__ = "tickets"

    id: Mapped[int] = mapped_column(primary_key=True)
    price: Mapped[float] = mapped_column(nullable=True)
    show: Mapped[str | None]
    user: Mapped[str | None]
```

We just created a `Ticket` class that will be used to match the `tickets` table into our SQL database.

Creating abstraction layers

In SQLAlchemy, the *engine* manages database connections and executes SQL statements, while a *session* allows querying, inserting, updating, and deleting data within a transactional context, ensuring consistency and atomicity. Sessions are bound to an engine for communication with the database.

We will start by creating a function that returns the engine. In a new module called db_connection.py, under the `app` folder, let's write the function as follows:

```
from sqlalchemy.ext.asyncio import (
    create_async_engine,
)
from sqlalchemy.orm import sessionmaker

SQLALCHEMY_DATABASE_URL = (
    "sqlite+aiosqlite:///.database.db"
)


def get_engine():
    return create_async_engine(
        SQLALCHEMY_DATABASE_URL, echo=True
    )
```

You may have observed that the SQLALCHEMY_DATABASE_URL database URL uses the `sqlite` and `aiosqlite` modules.

This implies that we will use an SQLite database where the operations will happen via the `aiosqlite` asynchronous library that supports the `asyncio` library.

Then, we will use a session maker to specify that the session will be asynchronous, as follows:

```
from sqlalchemy.ext.asyncio import (
    AsyncSession,
)

AsyncSessionLocal = sessionmaker(
    autocommit=False,
    autoflush=False,
    bind=get_engine(),
    class_=AsyncSession,
)
```

```
async def get_db_session():
    async with AsyncSessionLocal() as session:
        yield session
```

The get_db_session function will be used as a dependency for each endpoint interacting with the database.

Initializing a database connection

Once we have the abstraction layers, we need to create our FastAPI server object and start the database classes when the server runs. We can do it in the main.py module under the app folder:

```
from contextlib import asynccontextmanager

from fastapi import FastAPI

from app.database import Base
from app.db_connection import (
    AsynSessionLocal,
    get_db_session
)

@asynccontextmanager
async def lifespan(app: FastAPI):
    engine = get_engine()
    async with engine.begin() as conn:
        await conn.run_sync(Base.metadata.create_all)
        yield
    await engine.dispose()

app = FastAPI(lifespan=lifespan)
```

To specify server actions at the startup event, we have used the lifespan parameter.

We have everything in place to connect our application with the database.

How it works...

The creation of the Ticket database mapping class tells our application how the database is structured, and the session will manage the transactions. Then, the engine will not only execute the operations but compare the mapping classes with the database, and it will create tables if any are missing.

To check that our app communicates with our database, let's spin up the server from the command line at the project root folder:

```
$ uvicorn app.main:app
```

You should see message logs on the command output that says table tickets have been created. Furthermore, open the .database.db file with the database reader you prefer, and the table should be there with the schema that is defined in the database.py module.

See also

You can see more about how to set up a database with SQLAlchemy and how to make it compatible with the asyncio module on the official documentation pages:

- *How to set up an SQLAlchemy database*: https://docs.sqlalchemy.org/en/20/orm/quickstart.html

- *SQLAlchemy* asyncio *extension reference*: https://docs.sqlalchemy.org/en/20/orm/extensions/asyncio.html

In this example, we have used an SQLite database by specifying the following:

```
SQLALCHEMY_DATABASE_URL = "sqlite+aiosqlite:///.database.db"
```

However, you can use SQLAlchemy to interact with multiple SQL databases such as **MySQL** or **PostgreSQL** by simply specifying the database driver, the asyncio-supported driver, and the database address.

For example, for MySQL, the connection string would look like this:

```
mysql+aiomysql://user:password@host:port/
dbname[?key=value&key=value...]
```

In this case, you need the aiomysql package installed in your environment.

You can check more on the official documentation pages:

- SQLAlchemy MySQL dialect: https://docs.sqlalchemy.org/en/20/dialects/mysql.html

- SQLAlchemy PostgreSQL dialect: https://docs.sqlalchemy.org/en/20/dialects/postgresql.html

Implementing CRUD operations

CRUD operations with a RESTful API can be implemented using HTTP methods (POST, GET, PUT, and DELETE) for web services. This recipe demonstrates how to use SQLAlchemy and asyncio to build CRUD operations asynchronously on an SQL database with the corresponding endpoints.

Getting ready

Before you start with the recipe, you need to have a database connection and a table in the dataset, as well as a matching class in the code base. If you completed the previous recipe, you should have them ready.

How to do it...

We'll begin by making an operations.py module under the app folder to contain our database operations by following these steps.

1. First, we can set up the operation to add a new ticket to the database as follows:

    ```python
    from sqlalchemy.ext.asyncio import AsyncSession
    from sqlalchemy.future import select

    from app.database import Ticket

    async def create_ticket(
        db_session: AsyncSession,
        show_name: str,
        user: str = None,
        price: float = None,
    ) -> int:
        ticket = Ticket(
            show=show_name,
            user=user,
            price=price,
        )

        async with db_session.begin():
            db_session.add(ticket)
            await db_session.flush()
            ticket_id = ticket.id
            await db_session.commit()
        return ticket_id
    ```

The function will give back the ID attached to the ticket when saved.

2. Then, let's create a function to get a ticket:

```python
async def get_ticket(
    db_session: AsyncSession, ticket_id: int
) -> Ticket | None:
    query = (
        select(Ticket)
        .where(Ticket.id == ticket_id)
    )
    async with db_session as session:
        tickets = await session.execute(query)
        return tickets.scalars().first()
```

If the ticket is not found, the function will return a None object.

3. Then, we build an operation to update only the price of the ticket:

```python
async def update_ticket_price(
    db_session: AsyncSession,
    ticket_id: int,
    new_price: float,
) -> bool:
    query = (
        update(Ticket)
        .where(Ticket.id == ticket_id)
        .values(price=new_price)
    )
    async with db_session as session:
        ticket_updated = await session.execute(query)
        await session.commit()
        if ticket_updated.rowcount == 0:
            return False
        return True
```

The function gives back False if the operation couldn't update any ticket.

4. To conclude the CRUD operations, we define a delete_ticket operation:

```python
async def delete_ticket(
    db_session: AsyncSession, ticket_id
) -> bool:
    async with db_session as session:
        tickets_removed = await session.execute(
```

```
                    delete(
                        Ticket
                    ).where(Ticket.id == ticket_id)
                )
                await session.commit()

                if tickets_removed.rowcount == 0:
                    return False
                return True
```

Similarly to the update operation, the function returns `False` if it does not find any ticket to delete.

5. After defining the operations, we can expose them by creating the corresponding endpoints in the `main.py` module.

Let's do it for the create operation right after defining the app server:

```python
from typing import Annotated

from sqlalchemy.ext.asyncio import AsyncSession
from app.db_connection import (
    AsyncSessionLocal,
    get_engine,
    get_session
)
from app.operations import create_ticket

# rest of the code

class TicketRequest(BaseModel):
    price: float | None
    show: str | None
    user: str | None = None


@app.post("/ticket", response_model=dict[str, int])
async def create_ticket_route(
    ticket: TicketRequest,
    db_session: Annotated[
        AsyncSession,
        Depends(get_db_session)
    ]
):
    ticket_id = await create_ticket(
```

```
        db_session,
        ticket.show,
        ticket.user,
        ticket.price,
    )
    return {"ticket_id": ticket_id}
```

The remaining operations can be exposed in the same way.

Exercise

Similarly to what we did for the `create_ticket` operation, expose the other operations (get, update, and delete) with the respective endpoints.

How it works...

The functions created to interact with the database are exposed through the endpoints. This means that an external user will execute the operations by calling the respective endpoints.

Let's verify that the endpoint works correctly.

Start the server from the command line as usual by running the following:

```
$ uvicorn app.main:app
```

Then, go to the interactive documentation link at `http://localhost:8000/docs`, and you will see the endpoints you just created. Experiment with them in different combinations and see the results in the `.database.db` database file.

You have just created CRUD operations to interact with an SQL database by using `sqlalchemy` with the `asyncio` library.

Exercise

Make a `tests` folder in the root project folder and write all the unit tests for the operation functions and the endpoints. You can refer to *Chapter 5, Testing and Debugging FastAPI Applications*, to learn how to unit test FastAPI applications.

Working with migrations

Database migrations let you version control your database schema and keep it consistent across environments. They also help you automate the deployment of your database changes and track the history of your schema evolution.

The recipe shows you how to use **Alembic**, a popular tool for managing database migrations in Python. You will learn how to create, run, and roll back migrations and how to integrate them with your ticketing system.

Getting ready

To use the recipe, you need to have `alembic` in your environment. You can install it with `pip`, if you didn't do it with the `requirements.txt` file from the GitHub repository, by typing this on the command line:

```
$ pip install alembic
```

You also need to make sure you have at least one class that corresponds to the table in the database you want to create. If you don't have one, go back to the *Setting up SQLAlchemy* recipe and make one. If you're already running the application, delete the `.database.db` file that the application has created.

How to do it...

To configure Alembic and manage database migrations, go through the following steps.

1. The first step is to set up `alembic`. In the project root folder, run the following command in the command line:

    ```
    $ alembic init alembic
    ```

    This command will make an `alembic.ini` file and an `alembic` folder with some files inside it. The `alembic.ini` file is a configuration file for `alembic`.

    If you copy the project from the GitHub repository make sure to delete the existing `alembic` folder before running the `alembic init` command.

2. Find the `sqlalchemy.url` variable and set the database URL to the following:

    ```
    sqlalchemy.url = sqlite:///.database.db
    ```

    This specifies that we are using an SQLite database If you use a different database, adjust it accordingly.

3. The `alembic` directory contains a folder version and an `env.py` file that has the variable for creating our database migrations.

    Open the `env.py` file and find the `target_metadata` variable. Set its value to the metadata of our application as follows:

    ```
    from app.database import Base

    target_metadata = Base.metadata
    ```

We can now create our first database migration script and apply the migration.

4. Execute the following command from the command line to create an initial migration:

```
$ alembic revision --autogenerate -m "Start database"
```

This will create a migration script automatically placed in the `alembic/versions` folder.

5. Make sure you removed the existing `.database.db` file, and let's execute our first migration with the following command:

```
$ alembic upgrade head
```

This will automatically rebuild the `.database.db` file with the `tickets` table in it.

How it works...

Once we have the first version of our database, let's see the migration in action.

Imagine we want to change the table in the `database.py` module while the application is already deployed in a production environment so that we can't delete any records when updating it.

Add some tickets to the database, then in the code, let's add a new field called `sold` that will indicate if the ticket has been sold or not:

```
class Ticket(Base):
    __tablename__ = "tickets"

    id: Mapped[int] = mapped_column(primary_key=True)
    price: Mapped[float] = mapped_column(nullable=True)
    show: Mapped[str | None]
    user: Mapped[str | None]
    sold: Mapped[bool] = mapped_column(default=False)
```

To make a new migration, run the following command:

```
$ alembic revision --autogenerate -m "Add sold field"
```

You will find a new script in the `alembic/versions` folder.

Run the migration command again:

```
$ alembic upgrade head
```

Open the database, and you will see that the `tickets` table schema has the `sold` field added to it, and no record has been deleted.

You just created a migration strategy that will seamlessly change our database while running without any data loss. From now on, remember to use migrations to track changes on database schemas.

See also

You can see more on how to manage database migrations with Alembic at the official documentation links:

- *Setting up Alembic*: `https://alembic.sqlalchemy.org/en/latest/tutorial.html`

- *Autogenerating migrations*: `https://alembic.sqlalchemy.org/en/latest/autogenerate.html`

Handling relationships in SQL databases

Database relationships are associations between two or more tables that allow you to model complex data structures and perform queries across multiple tables. In this recipe, you will learn how to implement one-to-one, many-to-one, and many-to-many relationships for the existing ticketing system application. You will also see how to use SQLAlchemy to define your database schema relationships and query the database.

Getting ready

To follow the recipe, you need to have the core of the application already implemented with at least one table in it. If you have already done that, you will also have the necessary packages ready. We will keep on working on our ticketing system platform application.

How to do it...

We will now proceed to set up relationships. We will show an example for each type of SQL table relationship.

One to one

We will demonstrate the one-to-one relationship by making a new table that holds details about the ticket.

One-to-one relationships are used to group specific information about a record in a separate logic.

That being said, let's make the table in the `database.py` module. The records will have information such as the seat associated with the ticket, with a ticket type that we will use as a label for possible information. Let's create the table in two steps.

1. First, we will add the ticket details reference to the existing `Ticket` class:

    ```
    class Ticket(Base):
        __tablename__ = "tickets"

        id: Mapped[int] = mapped_column(primary_key=True)
    ```

```
price: Mapped[float] = mapped_column(
    nullable=True
)
show: Mapped[str | None]
user: Mapped[str | None]
sold: Mapped[bool] = mapped_column(default=False)
details: Mapped["TicketDetails"] = relationship(
    back_populates="ticket"
)
```

2. Then, we create the table to map the ticket's details as follows:

```
from sqlalchemy import ForeignKey


class TicketDetails(Base):
    __tablename__ = "ticket_details"

    id: Mapped[int] = mapped_column(primary_key=True)
    ticket_id: Mapped[int] = mapped_column(
        ForeignKey("tickets.id")
    )
    ticket: Mapped["Ticket"] = relationship(
        back_populates="details"
    )
    seat: Mapped[str | None]
    ticket_type: Mapped[str | None]
```

Once the database classes are set up to accommodate the new table, we can proceed to update the CRUD operations with the following steps.

1. To update ticket details, let's create a dedicated function in the `operations.py` module:

```
async def update_ticket_details(
    db_session: AsyncSession,
    ticket_id: int,
    updating_ticket_details: dict,
) -> bool:
    ticket_query = update(TicketDetails).where(
        TicketDetails.ticket_id == ticket_id
    )

    if updating_ticket_details != {}:
        ticket_query = ticket_query.values(
```

```
                *updating_ticket_details
        )

        result = await db_session.execute(
            ticket_query
        )
        await db_session.commit()
        if result.rowcount == 0:
            return False

    return True
```

The function will return `False` if no records have been updated.

2. Next, modify the `create_ticket` function to consider the details of the ticket and create an endpoint to expose the updating operation we just created, like so:

```
async def create_ticket(
    db_session: AsyncSession,
    show_name: str,
    user: str = None,
    price: float = None,
) -> int:
    ticket = Ticket(
        show=show_name,
        user=user,
        price=price,
        details=TicketDetails(),
    )

    async with db_session.begin():
        db_session.add(ticket)
        await db_session.flush()
        ticket_id = ticket.id
        await db_session.commit()
    return ticket_id
```

In this example, each time a ticket is created, an empty record of ticket details is created as well to keep the database consistent.

This was the minimum setup to handle one-to-one relationships. We will continue by setting up many-to-one relationships.

Many to one

A ticket can be associated with an event, and an event can have multiple tickets. To showcase a many-to-one relationship, we will create an events table that will have a relationship with the tickets table. Let's go through the following steps:

Let's first create a column in the tickets table that will accommodate the reference to the events table in the database.py module, as follows:

```
class Ticket(Base):
    __tablename__ = "tickets"
    # skip existing columns
    event_id: Mapped[int | None] = mapped_column(
        ForeignKey("events.id")
    )
    event: Mapped["Event | None"] = relationship(
        back_populates="tickets"
    )
```

Then, we create an Event class to map the events table into the database:

```
class Event(Base):
    __tablename__ = "events"

    id: Mapped[int] = mapped_column(primary_key=True)
    name: Mapped[str]
    tickets: Mapped[list["Ticket"]] = relationship(
        back_populates="event"
    )
```

ForeignKey, in this case, is defined only in the Ticket class since the event associated can be only one.

This is all you need to create a many-to-one relationship.

> **Exercise**
> You can add to the application the operations to create an event and specify the number of tickets to create with it. Once you've done this, expose the operation with the corresponding endpoint.

Many to many

Let's imagine that we have a list of sponsors that can sponsor our events. Since we can have multiple sponsors that can sponsor multiple events, this situation is best representative of a many-to-many relationship.

To work with many-to-many relationships, we need to define a class for the concerned tables and another class to track the so-called *association table*.

Let's start by defining a column to accommodate relationships in the Event class:

```python
class Event(Base):
    __tablename__ = "events"

    # existing columns
    sponsors: Mapped[list["Sponsor"]] = relationship(
        secondary="sponsorships",
        back_populates="events",
    )
```

Then, we can create a class to map the sponsors table:

```python
class Sponsor(Base):
    __tablename__ = "sponsors"

    id: Mapped[int] = mapped_column(primary_key=True)
    name: Mapped[str] = mapped_column(unique=True)
    events: Mapped[list["Event"]] = relationship(
        secondary="sponsorships",
        back_populates="sponsors",
    )
```

As you might have noticed, the class contains columns to accommodate the events reference.

Finally, we can define an association table that will be the sponsorships table:

```python
class Sponsorship(Base):
    __tablename__ = "sponsorships"

    event_id: Mapped[int] = mapped_column(
        ForeignKey("events.id"), primary_key=True
    )
    sponsor_id: Mapped[int] = mapped_column(
        ForeignKey("sponsors.id"), primary_key=True
    )
    amount: Mapped[float] = mapped_column(
        nullable=False, default=10
    )
```

The association table can contain information on the relationship itself. For example, in our case, a piece of useful information is the amount provided by the sponsor for the event.

This is all you need to create many-to-many relationships for your ticketing system platform.

Exercise

To complete your application, create an operations function with the relative endpoints to do the following:

- Add a sponsor to the database.

- Add a sponsorship with the amount. If the sponsorship already exists, replace the sponsorship with the new amount.

See also

You can dive deeper into handling relationships with SQLAlchemy at the following official documentation page:

- *SQLAlchemy basic relationships*: `https://docs.sqlalchemy.org/en/20/orm/basic_relationships.html`

Optimizing SQL queries for performance

Optimizing SQL queries is key in database management, as it enhances efficiency, scalability, cost-effectiveness, user satisfaction, data integrity, compliance, and security.

This recipe shows how to make applications run faster by improving SQL queries. Queries that use fewer resources and less time can enhance user satisfaction and application capacity. Improving SQL queries is a repeated process, but you could use some tips that could assist you.

Getting ready

Make sure you have an existing application running using SQLAlchemy for database interaction or to keep working on the ticketing system application all along the chapter. Also, basic knowledge of SQL and database schema design can be beneficial.

How to do it...

Improving SQL queries is a process that involves several steps. As with most optimization processes, many steps are specific to the use case, but there are general rules that can help optimize SQL queries overall, such as the following:

- Avoid *N+1* queries
- Use the JOIN statement sparingly
- Minimize data to fetch

We will apply each with a significant example.

Avoiding N+1 queries

The N+1 query issue happens when your application does one query to get a list of items and then loops over those items to get related data, making N more queries.

Let's say we want an endpoint to show all events with the associated sponsors. A first try might be to fetch the events table and loop over the events to fetch the sponsors table. This solution means a first query to get the events and N more queries to get the sponsors for each event, which is exactly what we want to avoid.

The solution is to load all related records in the query to retrieve the related sponsors. This is technically called *eager loading*.

In SQLAlchemy, this is done by using a joinedload option so that the function operation will look like this:

```
async def get_events_with_sponsors(
    db_session: AsyncSession
) -> list[Event]:
    query = (
        select(Event)
        .options(
            joinedload(Event.sponsors)
        )
    )
    async with db_session as session:
        result = await session.execute(query)
        events = result.scalars().all()

    return events
```

The `joinedload` method will include a `JOIN` operation on the query, so it is no longer necessary to make N queries to get the sponsors.

Using the join statement sparingly

Joined tables can make the query easier to read. But be careful and only join tables that you need for your query.

Suppose we want to get a list of sponsors names with the amount given for a certain event in order from the highest to the lowest.

We can use multiple joins since we need to fetch three tables. The function would look like this:

```
async def get_event_sponsorships_with_amount(
    db_session: AsyncSession, event_id: int
):
    query = (
        select(Sponsor.name, Sponsorship.amount)
        .join(
            Sponsorship,
            Sponsorship.sponsor_id == Sponsor.id,
        )
        .join(
            Event,
            Sponsorship.event_id == Event.id
        )
        .order_by(Sponsorship.amount.desc())
    )
    async with db_session as session:
        result = await session.execute(query)
        sponsor_contributions = result.fetchall()
    return sponsor_contributions
```

The double join implies to call the `events` table that we won't use, so it would be much more efficient to organize the query as follows:

```
async def get_event_sponsorships_with_amount(
    db_session: AsyncSession, event_id: int
):
    query = (
        select(Sponsor.name, Sponsorship.amount)
        .join(
            Sponsorship,
            Sponsorship.sponsor_id == Sponsor.id,
```

```
        )
        .where(Sponsorship.event_id == event_id)
        .order_by(Sponsorship.amount.desc())
    )
    async with db_session as session:
        result = await session.execute(query)
        sponsor_contributions = result.fetchall()
    return sponsor_contributions
```

This will return what we need without selecting the events table at all.

Minimizing data to fetch

Fetching more data than needed can slow down your queries and the application.

Use SQLAlchemy's load_only function to load only specific columns from the database.

Imagine that for a marketing analysis, we are asked to make a function that gets a list of tickets with only the ticket ID, the user, and the price:

```
async def get_events_tickets_with_user_price(
    db_session: AsyncSession, event_id: int
) -> list[Ticket]:
    query = (
        select(Ticket)
        .where(Ticket.event_id == event_id)
        .options(
            load_only(
                Ticket.id, Ticket.user, Ticket.price
            )
        )
    )
    async with db_session as session:
        result = await session.execute(query)
        tickets = result.scalars().all()
    return tickets
```

We now try to retrieve the tickets from this function, as follows:

```
tickets = await get_events_tickets_with_user_price(
    session, event_id
)
```

You will notice that each element only has the `id`, `user`, and `price` fields and it will give an error if you attempt to access the `show` field, for example. In larger applications, this can reduce memory usage and make responses much faster.

There's more...

SQL query optimization involves more than what the recipe showed. Often, choosing a certain SQL database depends on specific optimization needs.

Different SQL databases may have different strengths and weaknesses in handling these factors, depending on their architecture and features. For example, some SQL databases may support partitioning, sharding, replication, or distributed processing, which can improve the scalability and availability of data. Some SQL databases may offer more advanced query optimization techniques, such as cost-based optimization, query rewriting, or query caching, which can reduce the execution time and resource consumption of queries. Some SQL databases may implement different storage engines, transaction models, or index types, which can affect the performance and consistency of data operations.

Therefore, when choosing an SQL database for a specific application, it is important to consider the characteristics and requirements of the data and queries, and compare the capabilities and limitations of the available SQL databases. A good way to do this is to benchmark the performance of SQL databases using realistic datasets and queries and measure the relevant metrics, such as throughput, latency, accuracy, and reliability. By doing so, one can find the optimal SQL database for the given scenario and also identify potential areas for improvement in the database design and query formulation.

Securing sensitive data in SQL databases

Sensitive data, such as personal information, financial records, or confidential documents, is often stored in SQL databases for various applications and purposes. However, this also exposes the data to potential risks of unauthorized access, theft, leakage, or corruption. Therefore, it is essential to secure sensitive data in SQL databases and protect it from malicious attacks or accidental errors.

This recipe will show how to store sensitive data, such as credit card information, in SQL databases.

Getting ready

To follow the recipe, you need to have an application with a database connection already in place.

Furthermore, we will use the `cryptography` package. If you haven't installed it with the `requirements.txt` file, you can do it by running this command in your environment:

```
$ pip install cryptography
```

A sound knowledge of cryptography can be beneficial but is not necessary.

How to do it...

We will make a new table from the ground up to store credit card information. Some of the information, such as credit card numbers and **Card Verification Values** (**CVV**), will not be saved in clear text in our database but rather encrypted. Since we need to get it back, we will use a symmetric encryption that needs a key. Let's make the process through the following steps.

1. Let's start by creating a class in the `database.py` module that corresponds to the `credit_card` table in our database, as follows:

    ```
    class CreditCard(Base):
        __tablename__ = "credit_cards"

        id: Mapped[int] = mapped_column(primary_key=True)
        number: Mapped[str]
        expiration_date: Mapped[str]
        cvv: Mapped[str]
        card_holder_name: Mapped[str]
    ```

2. Next, in the `app` folder, we create a module named `security.py` where we will write our code for encrypting and decrypting data using **Fernet symmetric encryption**, as follows:

    ```
    from cryptography.fernet import Fernet


    cypher_key = Fernet.generate_key()
    cypher_suite = Fernet(cypher_key)
    ```

    The `cypher_suite` object will be used to define the encryption and decryption function.

    It is worth mentioning that in a production environment, the `cypher_key` object can be either kept in an external service that offers rotation or created at startup, based on the security needs of the business.

3. In the same module, we can create a function to encrypt credit card info and one to decrypt it as follows:

    ```
    def encrypt_credit_card_info(card_info: str) -> str:
        return cypher_suite.encrypt(
            card_info.encode()
        ).decode()


    def decrypt_credit_card_info(
        encrypted_card_info: str,
    ) -> str:
    ```

```
        return cypher_suite.decrypt(
            encrypted_card_info.encode()
        ).decode()
```

Those functions will be used when writing and reading from the database.

4.   Then, we can write a storing operation in the same `security.py` module as follows:

```
from sqlalchemy import select
from sqlalchemy.ext.asyncio import AsyncSession

from app.database import CreditCard

async def store_credit_card_info(
    db_session: AsyncSession,
    card_number: str,
    card_holder_name: str,
    expiration_date: str,
    cvv: str,
):
    encrypted_card_number = encrypt_credit_card_info(
        card_number
    )
    encrypted_cvv = encrypt_credit_card_info(cvv)

    # Store encrypted credit card information
    # in the database
    credit_card = CreditCard(
        number=encrypted_card_number,
        card_holder_name=card_holder_name,
        expiration_date=expiration_date,
        cvv=encrypted_cvv,
    )

    async with db_session.begin():
        db_session.add(credit_card)
        await db_session.flush()
        credit_card_id = credit_card.id
        await db_session.commit()
    return credit_card_id
```

Each time the function is awaited, the credit card information will be stored with the confidential data encrypted.

5. Similarly, we can define a function to retrieve the encrypted credit card information from the database as follows:

```
async def retrieve_credit_card_info(
    db_session: AsyncSession, credit_card_id: int
):
    query = select(CreditCard).where(
        CreditCard.id == credit_card_id
    )

    async with db_session as session:
        result = await session.execute(query)
        credit_card = result.scalars().first()

    credit_card_number = decrypt_credit_card_info(
            credit_card.number
        ),
    cvv = decrypt_credit_card_info(credit_card.cvv)
    card_holder = credit_card.card_holder_name
    expiry = credit_card.expiration_date

    return {
        "card_number": credit_card_number,
        "card_holder_name": card_holder,
        "expiration_date": expiry,
        "cvv": cvv
    }
```

We have just developed code to save confidential information in our database.

Exercise

We just saw the backbone of how to store sensitive data securely. You can complete the feature by yourself by doing the following:

- Writing unit tests for our encryption operations. In the `tests` folder, let's create a new test module called `test_security.py`. Verify that the credit card is securely saved in our database, but the fields for credit card numbers and CVV are encrypted.

- Creating endpoints to store, retrieve, and delete credit card information in the database.

- Associating a credit card with a sponsor and managing the relative CRUD operations.

See also

We have used Fernet symmetric encryption to encrypt credit card information. You can have a deeper insight about it at the following link:

- *Fernet symmetric encryption*: `https://cryptography.io/en/latest/fernet/`

Handling transactions and concurrency

In the realm of database management, two critical aspects govern the reliability and performance of applications: handling transactions and managing concurrency.

Transactions, encapsulating a series of database operations, are fundamental for maintaining data consistency by ensuring that changes occur as a single unit of work. Concurrency, on the other hand, addresses the challenge of managing simultaneous access to shared resources by multiple users or processes.

The relationship between transactions and concurrency becomes apparent when considering scenarios where multiple transactions may attempt to access or modify the same data concurrently. Without proper concurrency control mechanisms such as locking, transactions could interfere with each other, potentially leading to data corruption or inconsistencies.

The recipe will show how to manage transactions and concurrency with FastAPI and SQLAlchemy by emulating the process of selling tickets from the ticketing platform we created.

Getting ready

You need a CRUD application as the basis for the recipe, or you can continue to use the ticketing system application that we have been using throughout the chapter.

How to do it...

The most significant situation where transaction and concurrency become important is in managing updating operations, such as with the sales ticket for our application.

We will begin by creating a function operation that will label our ticket as sold and give the name of the customer. Then, we will simulate two sales occurring at the same time and observe the outcome. To do so, follow these steps.

1. In the `operations.py` module, create the function to sell a ticket as follows::

```
async def sell_ticket_to_user(
    db_session: AsyncSession, ticket_id: int, user: str
) -> bool:
    ticket_query = (
```

```
                update(Ticket)
                .where(
                    and_(
                        Ticket.id == ticket_id,
                        Ticket.sold == False,
                    )
                )
                .values(user=user, sold=True)
        )

        async with db_session as session:
            result = (
                await db_session.execute(ticket_query)
            )
            await db_session.commit()
            if result.rowcount == 0:
                return False
        return True
```

The query will only sell the ticket if the ticket has not been sold yet; otherwise, the function will return `False`.

2. Let's try to add a ticket to our database and try to simulate two users buying the same ticket at the same time. Let's write all in the form of unit tests.

    We start by defining a fixture to write our ticket into the database in the `tests/conftest.py` file as follows:

```
@pytest.fixture
async def add_special_ticket(db_session_test):
    ticket = Ticket(
        id=1234,
        show="Special Show",
        details=TicketDetails(),
    )
    async with db_session_test.begin():
        db_session_test.add(ticket)
        await db_session_test.commit()
```

3. We can create a test by performing two concurrent sales with two separate database sessions (define another one as a different fixture) to do them at the same time in the `tests/test_operations.py` file:

```python
import asyncio

async def test_concurrent_ticket_sales(
    add_special_ticket,
    db_session_test,
    second_session_test,
):
    result = await asyncio.gather(
        sell_ticket_to_user(
            db_session_test, 1234, "Jake Fake"
        ),
        sell_ticket_to_user(
            second_session_test, 1234, "John Doe"
        ),
    )

    assert result in (
        [True, False],
        [False, True],
    )  # only one of the sales should be successful

    ticket = await get_ticket(db_session_test, 1234)

    # assert that the user who bought the ticket
    # correspond to the successful sale
    if result[0]:
        assert ticket.user == "Jake Fake"
    else:
        assert ticket.user == "John Doe"
```

In the test function, we run two coroutines at the same time by using the `asyncio.gather` function.

We just assume that only one user can purchase the ticket and they will match the successful transaction. Once we have created the test, we can execute with `pytest` as follows:

```
$ pytest tests/test_operations.py::test_concurrent_ticket_sales
```

The test will succeed, which means that the asynchronous session handles transaction conflicts.

> **Exercise**
>
> You have just created a draft of the selling ticket operation. As an exercise, you can improve the draft by doing the following:
>
> - Adding a table for users to the database
>
> - Adding the foreign key reference of the user on the ticket to make it sold
>
> - Creating an `alembic` migration for the database modification
>
> - Creating an API endpoint that exposes the `sell_ticket_to_user` function

There's more...

One of the fundamental challenges of database systems is to handle concurrent transactions from multiple users while preserving data consistency and integrity. Different types of transactions may have different requirements for how they access and modify data and how they deal with other transactions that may conflict with them. For example, a common way to manage concurrency is to use *locks*, which are mechanisms that prevent unauthorized or incompatible operations on data. However, locks can also introduce trade-offs between performance, availability, and correctness.

Depending on the business needs, some transactions may need to acquire locks for longer periods or at different levels of granularity, such as table-level or row-level. For example, SQLite only allows locks on a database level, while PostgreSQL allows locks till the row table level.

Another key aspect of managing concurrent transactions is the concept of *isolation levels*, which define the degree to which one transaction must be isolated from the effects of other concurrent transactions. Isolation levels ensure that transactions maintain data consistency despite simultaneous access and modification by multiple users.

The SQL standard defines four isolation levels, each offering different trade-offs between concurrency and data consistency:

1. **READ UNCOMMITTED**:

    - Transactions at this level allow dirty reads, meaning a transaction can see uncommitted changes made by other concurrent transactions.

    - Non-repeatable reads and phantom reads are possible.

    - This isolation level provides the highest concurrency but the lowest level of data consistency.

2. **READ COMMITTED**:

   - Transactions at this level only see changes committed by other transactions.

   - They do not allow dirty reads.

   - Non-repeatable reads are possible, but phantom reads can still occur.

   - This level strikes a balance between concurrency and consistency.

3. **REPEATABLE READ**:

   - Transactions at this level see a consistent snapshot of the data throughout the transaction.

   - Changes committed by other transactions after the transaction began are not visible.

   - Non-repeatable reads are prevented, but phantom reads can occur.

   - This level provides stronger consistency at the cost of some concurrency.

4. **SERIALIZABLE**:

   - Transactions at this level behave as if they are executed serially – that is, one after another.

   - They provide the highest level of data consistency.

   - Non-repeatable reads and phantom reads are prevented.

   - This level offers strong consistency but may result in reduced concurrency due to increased locking.

SQLite, for example, allows isolation, while MySQL and PostgreSQL offer all four transaction levels.

When the database supports it, in SQLAlchemy, you can set up the isolation level per engine or connection by specifying it as an argument when initializing.

For example, if you want to specify the isolation level at the engine level for PostgreSQL, the engine will be initialized as follows:

```
from sqlalchemy import create_engine
from sqlalchemy.orm import sessionmaker

eng = create_engine(
    "postgresql+psycopg2://scott:tiger@localhost/test",
    isolation_level="REPEATABLE READ",
)

Session = sessionmaker(eng)
```

All these choices in terms of locks and isolation level affect the architecture and design of the database system since not all SQL databases support it. Therefore, it is important to understand the principles and best practices of locking strategies and how they relate to the transaction behavior and the business logic.

You have just completed a comprehensive overview of integrating SQL databases with FastAPI. In the next chapter, we will explore integrating FastAPI applications with NoSQL databases.

See also

You can find more information about locking strategies for SQLite and PostgreSQL at the following links:

- *SQLite locking*: `https://www.sqlite.org/lockingv3.html`

- *PostgreSQL locking*: `https://www.postgresql.org/docs/current/explicit-locking.html`

Information on the isolation level for singular databases can be found on the respective documentation pages:

- *SQLite isolation*: `https://www.sqlite.org/isolation.html`

- *MySQL isolation levels*: `https://dev.mysql.com/doc/refman/8.0/en/innodb-transaction-isolation-levels.html`

- *PostgreSQL isolation levels*: `https://www.postgresql.org/docs/current/transaction-iso.html`

Also, a comprehensive guide on how to manage isolation levels with SQLAlchemy is available at the link:

- *SQLAlchemy session transaction*: `https://docs.sqlalchemy.org/en/20/orm/session_transaction.html`

7

Integrating FastAPI with NoSQL Databases

In this chapter, we will explore the integration of **FastAPI** with **NoSQL** databases. By crafting the backend of a music streaming platform application, you will learn how to set up and use **MongoDB**, a popular NoSQL database, with FastAPI.

You will also learn how to perform **create, read, update and delete** (**CRUD**) operations, work with indexes for performance optimization, and handle relationships in NoSQL databases. Additionally, you will learn how to integrate FastAPI with **Elasticsearch** for powerful search capabilities, secure sensitive data, and implement caching using **Redis**.

By the end of this chapter, you will have a solid understanding of how to effectively use NoSQL databases with FastAPI to improve the performance and functionality of your applications.

In this chapter, we're going to cover the following recipes:

- Setting up MongoDB with FastAPI
- CRUD operations in MongoDB
- Handling relationships in NoSQL databases
- Working with indexes in MongoDB
- Exposing sensitive data from NoSQL databases
- Integrating FastAPI with Elasticsearch
- Using Redis for caching in FastAPI

Technical requirements

To follow along with the recipes of the chapter, ensure your setup includes the following essentials:

- **Python**: A version 3.7 or higher should be installed on your computer
- **FastAPI**: Have the `fastapi` package in your working environment
- `asyncio`: Be familiar with the `asyncio` framework and `async`/`await` syntax since we will use it all along the recipes

The code used in the chapter is hosted on GitHub at this address: `https://github.com/PacktPublishing/FastAPI-Cookbook/tree/main/Chapter07`.

You can create a virtual environment for the project within the project root folder to manage dependencies efficiently and maintain project isolation.

Within your virtual environment, you can install all the dependencies at once by using `requirements.txt`, which is provided on the GitHub repository in the project folder:

```
$ pip install -r requirements.txt
```

General knowledge of the external tools we are going to use for each recipe can be beneficial, although not mandatory. Each recipe will provide you with a minimal explanation of the used tool.

Setting up MongoDB with FastAPI

In this recipe, you will learn how to set up MongoDB, a popular document-oriented NoSQL database, with FastAPI. You will learn how to manage Python packages to interact with MongoDB, create a database, and connect it to a FastAPI application. By the end of this recipe, you will have a solid understanding of how to integrate MongoDB with FastAPI to store and retrieve data for your applications.

Getting ready

To follow along with this recipe, you need Python and `fastapi package` installed in your environment.

Also, for this recipe, make sure you have a MongoDB instance running and reachable, and if not, set up a local one. Depending on your operating system and your personal preference, you can set up a local MongoDB instance in several ways. Feel free to consult the official documentation on how to install the community edition of MongoDB on your local machine at the following link: `https://www.mongodb.com/try/download/community`.

For the recipe and throughout the chapter, we will consider a local instance of MongoDB running on `http://localhost:27017`. If you run the MongoDB instance on a remote machine, or simply use a different port, adjust the URL reference accordingly.

You also need the `motor` package installed in your environment. If you haven't installed the packages with `requirements.txt`, you can install `motor` in your environment from the command line:

```
$ pip install motor
```

Motor is the asynchronous Python driver developed by **MongoDB Inc** and it allows Python code to interact with MongoDB through the `asyncio` library.

Once we have the MongoDB instance running and reachable and the `motor` package installed in your environment, we can proceed with the recipe.

How to do it...

Let's start by creating a project root folder called `streaming_platform` with an `app` subfolder. In `app`, we create a module called `db_connection.py`, which will contain the information on the connection with MongoDB.

Now, we will set up the connection through the following steps:

1. In the `db_connecion.py` module, let's define the MongoDB client:

    ```
    from motor.motor_asyncio import AsyncIOMotorClient

    mongo_client = AsyncIOMotorClient(
        "mongodb://localhost:27017"
    )
    ```

    We will use the `mongo_client` object each time we need to interact with the MongoDB instance that is running at `http://localhost:27017`.

2. In the `db_connection.py` module, we will create a function to ping the MongoDB instance to ensure it is running. But first, we retrieve the `uvicorn` logger, used by the FastAPI server, to print messages to the terminal:

    ```
    import logging

    logger = logging.getLogger("uvicorn.error")
    ```

3. Then, let's create the function to ping the MongoDB as follows:

    ```
    async def ping_mongo_db_server():
        try:
            await mongo_client.admin.command("ping")
            logger.info("Connected to MongoDB")
        except Exception as e:
    ```

```
        logger.error(
            f"Error connecting to MongoDB: {e}"
        )
        raise e
```

The function will ping the server, and if it doesn't receive any response, it will propagate an error that will stop the code from running.

4. Finally, we need to run the `ping_mongo_db_server` function when starting the FastAPI server. In the `app` folder, let's create a `main.py` module with a context manager that will be used for the startup and shutdown of our FastAPI server:

```
from contextlib import asynccontextmanager
from app.db_connection import (
    ping_mongo_db_server,
)


@asynccontextmanager
async def lifespan(app: FastAPI):
    await ping_mongo_db_server(),
    yield
```

The `lifespan` context manager has to be passed as an argument to the `FastAPI` object:

```
from fastapi import FastAPI

app = FastAPI(lifespan=lifespan)
```

The server is wrapped in the `lifespan` context manager to execute the database check at startup.

To test it, make sure your MongoDB instance is already running and as usual, let's spin up the server from the command line:

```
$ uvicorn app.main:app
```

You will see the following log messages on the output:

```
INFO:    Started server process [1364]
INFO:    Waiting for application startup.
INFO:    Connected to MongoDB
INFO:    Application startup complete.
```

This message confirms that our application correctly communicates with the MongoDB instance.

You've just set up the connection between a FastAPI application and a MongoDB instance.

See also

You can see more on the Motor asynchronous driver on the MongoDB official documentation page:

- *Motor Async Driver Setup*: `https://www.mongodb.com/docs/drivers/motor/`

For startups and shutdown events of the FastAPI server, you can find more on this page:

- *FastAPI Lifespan Events*: `https://fastapi.tiangolo.com/advanced/events/`

CRUD operations in MongoDB

CRUD operations form the cornerstone of data manipulation in databases, enabling users to create, read, update, and delete data entities with efficiency, flexibility, and scalability.

This recipe will demonstrate how to create endpoints in FastAPI for creating, reading, updating, and deleting a document from a MongoDB database for the backbone of our streaming platform.

Getting ready

To follow along with the recipe, you need a database connection with MongoDB already in place with your application, otherwise, go to the previous recipe, *Setting up MongoDB with FastAPI*, which will show you in detail how to do it.

How to do it...

Before creating the endpoints for the CRUD operations, we have to initialize a database on the MongoDB instance for our streaming application.

Let's do it in a dedicated module in the `app` directory called `database.py` as follows:

```
from app.db_connection import mongo_client
database = mongo_client.beat_streaming
```

We've defined a database called `beat_streaming`, which will contain all the collections of our application.

On the MongoDB server side, we don't need any action to do since the `motor` library will automatically check for the existence of a database named `beat_streaming` and the eventual collections, and it will create them if they don't exist.

In the same module, we can create the function to return the database that will be used as a dependency in the endpoints for code maintainability:

```
def mongo_database():
    return database
```

Now, we can define our endpoints in main.py for each of the CRUD operations through the following steps.

1. Let's start by creating the endpoint to add a song to the songs collection:

    ```
    from bson import ObjectId

    from fastapi import Body, Depends
    from app.database import mongo_database
    from fastapi.encoders import ENCODERS_BY_TYPE

    ENCODERS_BY_TYPE[ObjectId] = str

    @app.post("/song")
    async def add_song(
        song: dict = Body(
            example={
                "title": "My Song",
                "artist": "My Artist",
                "genre": "My Genre",
            },
        ),
        mongo_db=Depends(mongo_database),
    ):
        await mongo_db.songs.insert_one(song)

        return {
            "message": "Song added successfully",
            "id": song["_id"],
        }
    ```

The endpoint takes a general JSON in the body and returns the ID affected from the database. The ENCONDERS_BY_TYPE[ObjectID] = str line specifies to the FastAPI server that the song["_id"] document ID has to be decoded as a string.

One of the reasons to choose a NoSQL database is the freedom from SQL schema, which allows for more flexibility in managing data. However, it can be helpful to provide an example to follow in the documentation. This is achieved by using the `Body` object class with the example parameter.

2. The endpoint to retrieve a song will be quite straightforward:

```
@app.get("/song/{song_id}")
async def get_song(
    song_id: str,
    db=Depends(mongo_database),
):
    song = await db.songs.find_one(
        {
            "_id": ObjectId(song_id)
            if ObjectId.is_valid(song_id)
            else None
        }
    )
    if not song:
        raise HTTPException(
            status_code=404,
            detail="Song not found"
        )
    return song
```

The application will search for a song with the specified ID and return a `404` error if none is found.

3. To update a song, the endpoint will look like this:

```
@app.put("/song/{song_id}")
async def update_song(
    song_id: str,
    updated_song: dict,
    db=Depends(mongo_database),
):
    result = await db.songs.update_one(
        {
            "_id": ObjectId(song_id)
            if ObjectId.is_valid(song_id)
            else None
        },
        {"$set": updated_song},
    )
```

```
    if result.modified_count == 1:
      return {
          "message": "Song updated successfully"
      }

    raise HTTPException(
        status_code=404, detail="Song not found"
    )
```

The endpoint will return a 404 error if the song id does not exist, otherwise it will update only the fields specified in the body request.

4. Finally, the delete operation endpoint can be done as follows:

```
@app.delete("/song/{song_id}")
async def delete_song(
    song_id: str,
    db=Depends(mongo_database),
):
    result = await db.songs.delete_one(
        {
            "_id": ObjectId(song_id)
            if ObjectId.is_valid(song_id)
            else None
        }
    )
    if result.deleted_count == 1:
        return {
            "message": "Song deleted successfully"
        }

    raise HTTPException(
        status_code=404, detail="Song not found"
    )
```

You have just created the endpoints to interact with a MongoDB database.

Now, spin up the server from the command line and test the endpoints you just created from the interactive documentation at http://localhost:8000/docs.

If you follow along with the GitHub repository, you can also prefill the database with the script `fill_mongo_db_database.py` at the link: `https://github.com/PacktPublishing/FastAPI-Cookbook/blob/main/Chapter07/streaming_platform/fill_mongo_db_database.py`

Make sure you download also the `songs_info.py` in the same folder.

You can then run the script from the terminal as follows:

```
$ python fill_mongo_db_database.py
```

If you call the endpoint `GET /songs` you will have a long list of songs pre filled to test your API.

See also

You can investigate the operations provided by `motor` to interact with a MongoDB instance further at the official documentation link:

- *Motor MongoDB Aynscio Tutorial*: `https://motor.readthedocs.io/en/stable/tutorial-asyncio.html`

Handling relationships in NoSQL databases

Unlike relational databases, NoSQL databases do not support joins or foreign keys for defining relationships between collections.

Schema-less databases, such as MongoDB, do not enforce relationships like traditional relational databases. Instead, two primary approaches can be used for handling relationships: **embedding** and **referencing**.

Embedding involves storing related data within a single document. This approach is suitable for all types of relationships, provided that the embedded data is closely tied to the parent document. This technique is good for read performance for frequently accessed data and atomic updates with a single document. However, it can easily lead to size limitation problems with data duplication and potential inconsistencies if the embedded data changes frequently.

Referencing involves storing references to related documents using their object ID or other unique identifiers. This approach is suitable for many-to-one and many-to-many relationships where the related data is huge and is shared across multiple documents.

This technique reduces data duplication and improves flexibility to update related data independently, but, on the other hand, increases the complexity of reading operations due to multiple queries leading to slower performances when fetching related data.

In this recipe, we'll explore both techniques for handling relationships between data entities in MongoDB by adding new collections to our streaming platform and making them interact.

Getting ready

We will continue building our streaming platform. Make sure you have followed all the previous recipes in this chapter, or you can apply the steps to an existing application that interacts with a NoSQL database.

How to do it...

Let's see how to implement relationships for both embedding and referencing techniques.

Embedding

A suitable candidate to showcase embedded relationships for songs is a collection of albums. Album information does not change often, if not never, once it is published.

The `album` document will embedded into the `song` document with a nested field:

```
{
    "title": "Title of the Song",
    "artist": "Singer Name",
    "genre": "Music genre",
    "album": {
        "title": "Album Title",
        "release_year": 2017,
    },
}
```

When using MongoDB, we can retrieve information about an album and a song using the same endpoint. This means that when we create a new song, we can directly add information about the album it belongs to. We specify the way we want the document song to be stored, and MongoDB takes care of the rest.

Spin up the server and test the POST /song endpoint. In the JSON body, include information about the album. Take note of the ID retrieved and use it to call the GET /song endpoint. Since we haven't defined any response schema restriction in the response model, the endpoint will return all the document information retrieved from the database including the album.

For this use case example, there is nothing to worry about, but for some applications, you might not want to disclose a field to the end user. You can either define a response model (see *Chapter 1, First Steps with FastAPI*, in the *Defining and using request and response models* recipe) or drop the field from the `dict` object before it is returned.

You have just defined a many-to-one relationship with the embedding strategy that relates songs to albums.

Referencing

A typical use case for referencing relationships can be the creation of a playlist. A playlist contains multiple songs, and each song can appear in different playlists. Furthermore, playlists are often changed or updated, so it respond to the need for a referencing strategy to manage relationships.

On the database side, we don't need any action so we will directly proceed to create the endpoint to create the playlist and the one to retrieve the playlist with all song information.

1. You can define the endpoint to create a playlist in the `main.py` module:

```python
class Playlist(BaseModel):
    name: str
    songs: list[str] = []


@app.post("/playlist")
async def create_playlist(
    playlist: Playlist = Body(
        example={
            "name": "My Playlist",
            "songs": ["song_id"],
        }
    ),
    db=Depends(mongo_database),
):
    result = await db.playlists.insert_one(
        playlist.model_dump()
    )
    return {
        "message": "Playlist created successfully",
        "id": str(result.inserted_id),
    }
```

The endpoint requires a JSON body specifying the playlist name and the list of song IDs to include, and it returns the playlist ID.

2. The endpoint to retrieve the playlist will take as an argument the playlist ID. You can code it as follows:

```python
@app.get("/playlist/{playlist_id}")
async def get_playlist(
    playlist_id: str,
    db=Depends(mongo_database),
):
```

```
playlist = await db.playlists.find_one(
    {
        "_id": ObjectId(playlist_id)
        if ObjectId.is_valid(playlist_id)
        else None
    }
)
if not playlist:
    raise HTTPException(
        status_code=404,
        detail="Playlist not found"
    )

songs = await db.songs.find(
    {
        "_id": {
            "$in": [
                ObjectId(song_id)
                for song_id in playlist["songs"]
            ]
        }
    }
).to_list(None)

return {
    "name": playlist["name"],
    "songs": songs
}
```

Notice that the song IDs in the playlist collection are stored as strings, not `ObjectId`, which means that they have to be converted when queried.

Also, to receive the list of songs for the playlist, we had to make two queries: one for the playlist and one to retrieve the songs based on their IDs.

Now that you build the endpoints to create and retrieve playlists, spin up the server:

```
$ uvicorn app.main:app
```

Go to the interactive documentation at `http://localhost:8000/docs` and you will see the new endpoints: `POST /playlist` and `GET /playlist`.

To test the endpoints, create some songs and note their IDs. Then, create a playlist and retrieve the playlist with the `GET /playlist` endpoint. You will see that the response will contain the songs with all the information including the album.

At this point, you have all the tools to manage relationships between collections in MongoDB.

See also

We just saw how to manage relationships with MongoDB and create relative endpoints. Feel free to check the official MongoDB guidelines at this link:

- *MongoDB Model Relationships*: `https://www.mongodb.com/docs/manual/applications/data-models-relationships/`

Working with indexes in MongoDB

An **index** is a data structure that provides a quick lookup mechanism for locating specific pieces of data within a vast dataset. Indexes are crucial for enhancing query performance by enabling the database to quickly locate documents based on specific fields.

By creating appropriate indexes, you can significantly reduce the time taken to execute queries, especially for large collections. Indexes also facilitate the enforcement of uniqueness constraints and support the execution of sorted queries and text search queries.

In this recipe, we'll explore the concept of indexes in MongoDB and we will create indexes to improve search performances for songs in our streaming platform.

Getting ready

To follow along with the recipe, you need to have a MongoDB instance already set up with at least a collection to apply indexes. If you are following along with the cookbook, make sure you went through the *Setting up MongoDB with FastAPI* and *CRUD operations in MongoDB* recipes.

How to do it...

Let's imagine we need to search for songs released in a certain year. We can create a dedicated endpoint directly in the `main.py` module as follows:

```python
@app.get("/songs/year")
async def get_songs_by_released_year(
    year: int,
    db=Depends(mongo_database),
):
    query = db.songs.find({"album.release_year": year})

    songs = await query.to_list(None)
    return songs
```

The query will fetch all documents and filter the one with a certain `release_year`. To speed up the query, we can create a dedicated index on the release year. We can do it at the server startup in the `lifespan` context manager in `main.py`. A text search in MongoDB won't be possible without a text index.

First, at the startup server, let's create a text index based on the `artist` field of the collection document. To do this, let's modify the `lifespan` context manager in the `main.py` module:

```
@asynccontextmanager
async def lifespan(app: FastAPI):
    await ping_mongo_db_server(),
    db = mongo_database()
    await db.songs.create_index({"album.release_year": -1})
    yield
```

The `create_index` method will create an index based on the `release_year` field sorted in descending mode because of the `-1` value.

You've just created an index based on the `release_year` field.

How it works...

The index just created is automatically used by MongoDB when running the query.

Let's check it by leveraging the explain query method. Let's add the following log message to the endpoint to retrieve songs released in a certain year:

```
@app.get("/songs/year")
async def get_songs_by_released_year(
    year: int,
    db=Depends(mongo_database),
):
    query = db.songs.find({"album.release_year": year})
    explained_query = await query.explain()
    logger.info(
        "Index used: %s",
        explained_query.get("queryPlanner", {})
        .get("winningPlan", {})
        .get("inputStage", {})
        .get("indexName", "No index used"),
    )

    songs = await query.to_list(None)
    return songs
```

The `explained_query` variable holds information about the query such as the query execution or index used for the search.

If you run the server and call the `GET /songs/year` endpoint, you will see the following message log on the terminal output:

```
INFO:      Index used: album.release_year_-1
```

This confirms that the query has correctly used the index we created to run.

There's more...

Database indexes become necessary to run text search queries. Imagine we need to retrieve the songs of a certain artist.

To query and create the endpoint, we need to make a text index on the `artist` field. We can do it at the server startup like the previous index on `album.release_year`.

In the `lifespan` context manager, you can add the index creation:

```python
@asynccontextmanager
async def lifespan(app: FastAPI):
    await ping_mongodb_server(),

    db = mongo_database()
    await db.songs.drop_indexes()
    await db.songs.create_index({"release_year": -1})
    await db.songs.create_index({"artist": "text"})
    yield
```

Once we have created the index, we can proceed to create the endpoint to retrieve the song based on the artist's name.

In the same `main.py` module, create the endpoint as follows:

```python
@app.get("/songs/artist")
async def get_songs_by_artist(
    artist: str,
    db=Depends(mongo_database),
):
    query = db.songs.find(
        {"$text": {"$search": artist}}
    )
    explained_query = await query.explain()
    logger.info(
```

```
        "Index used: %s",
        explained_query.get("queryPlanner", {})
        .get("winningPlan", {})
        .get("indexName", "No index used"),
    )


    songs = await query.to_list(None)
    return songs
```

Spin up the server from the command line with the following:

```
$ uvicorn app.main:app
```

Go to the interactive documentation at `http:/localhost:8000/docs` and try to run the new `GET /songs/artist` endpoint.

Text searching allow you to fetch records based on text matching. If you have filled the database with the `fill_mongo_db_database.py` script you can try searching for Bruno Mars's songs by specifying the family name `"mars"`. The query will be:

```
http://localhost:8000/songs/artist?artist=mars
```

This will return at the least the song:

```
[
  {
    "_id": "667038acde3a00e55e764cf7",
    "title": "Uptown Funk",
    "artist": "Mark Ronson ft. Bruno Mars",
    "genre": "Funk/pop",
    "album": {
      "title": "Uptown Special",
      "release_year": 2014
    }
  }
]
```

Also, you will see a message on the terminal output like:

```
INFO:    Index used: artist_text
```

That means that the database has used the correct index to fetch the data.

> **Important note**
> By using the `explanation_query` variable, you can also check the difference in the execution time. However, you need a huge number of documents in your collection to appreciate the improvement.

See also

We saw how to build a text index for the search over the artist and a numbered index for the year of release. MongoDB allows you to do more, such as defining 2D sphere index types or compound indexes. Have a look at the documentation to discover the potential of indexing your MongoDB database:

- *Mongo Indexes*: `https://www.mongodb.com/docs/v5.3/indexes/`
- *MongoDB Text Search*: `https://www.mongodb.com/docs/manual/core/link-text-indexes/`

Exposing sensitive data from NoSQL databases

The way to expose sensitive data in NoSQL databases is pivotal to protecting sensitive information and maintaining the integrity of your application.

In this recipe, we will demonstrate how to securely view our data through database aggregations with the intent to expose it to a third-party consumer of our API. This technique is known as **data masking**. Then, we will explore some strategies and best practices for securing sensitive data in MongoDB and NoSQL databases in general.

By following best practices and staying informed about the latest security updates, you can effectively safeguard your MongoDB databases against potential security threats.

Getting ready

To follow the recipe, you need to have a running FastAPI application with a MongoDB connection already set up. If don't have it yet, have a look at the *Setting up MongoDB with FastAPI* recipe. In addition, you need a collection of sensitive data such as **Personal Identifiable Information** (**PII**) or other restricted information.

Alternatively, we can build a collection of users into our MongoDB database, `beat_streaming`. The document contains PIIs such as names and emails, as well as users actions on the platform. The document will look like this:

```
{
    "name": "John Doe",
    "email": "johndoe@email.com",
    "year_of_birth": 1990,
```

```
    "country":  "USA",
    "consent_to_share_data": True,
    "actions":  [
        {
            "action":  "basic subscription",
            "date":  "2021-01-01",
            "amount":  10,
        },

        {
            "action":  "unscription",
            "date":  "2021-05-01",
        },
    ],
}
```

The `consent_to_share_data` field stores the consent of the user to share behavioral data with third-party partners.

Let's first fill the collection users in our database. You can do this with a user's sample by running the script provided in the GitHub repository:

```
$ python fill_users_in_mongo.py
```

If everything runs smoothly, you should have the collection users in your MongoDB instance.

How to do it...

Imagine we need to expose users data for marketing research to a third-party API consumer for commercial purposes. The third-party consumer does not need PII information such as names or emails, and they are also not allowed to have data from users who didn't give their consent. This is a perfect use case to apply data masking.

In MongoDB, you can build aggregation pipelines in stages. We will do it step by step.

1.   Since the database scaffolding is an infrastructure operation rather than an application, let's create the pipeline with the view in a separate script that we will run separately from the server.

     In a new file called `create_aggregation_and_user_data_view.py`, let's start by defining the client:

```
from pymongo import MongoClient
client = MongoClient("mongodb://localhost:27017/")
```

Since we don't have any need to manage high traffic, we will use the simple `pymongo` client instead of the asynchronous one. We will reserve the asynchronous to the sole use of the application interactions.

2. The pipeline stage follows a specific aggregations framework. The first step of the pipeline will be to filter out the users who didn't approve the consent. This can be done with a `$redact` stage:

```
pipeline_redact = {
    "$redact": {
        "$cond": {
            "if": {
                "$eq": [
                    "$consent_to_share_data", True
                ]
            },
            "then": "$$KEEP",
            "else": "$$PRUNE",
        }
    }
}
```

3. Then, we filter out the emails that shouldn't be shared with a `$unset` stage:

```
pipeline_remove_email_and_name = {
    "$unset": ["email", "name"]
}
```

4. This part of the pipeline will prevent emails and names from appearing in the pipeline's output. We will split stage definition into three dictionaries for a better understanding.

    First, we define the action to obfuscate the day for each date:

```
obfuscate_day_of_date = {
    "$concat": [
        {
            "$substrCP": [
                "$$action.date",
                0,
                7,
            ]
        },
        "-XX",
    ]
}
```

5. Then, we map the new `date` field for each element of the actions list:

```
rebuild_actions_elements = {
    "input": "$actions",
    "as": "action",
    "in": {
        "$mergeObjects": [
            "$$action",
            {"date": obfuscate_day_of_date},
        ]
    },
}
```

6. Then, we use a `$set` operation to apply the `rebuild_actions_element` operation to every record like that:

```
pipeline_set_actions = {
    "$set": {
        "actions": {"$map": rebuild_actions_elements},
    }
}
```

7. Then, we gather the pipelines just created to define the entire pipeline stage:

```
pipeline = [
    pipeline_redact,
    pipeline_remove_email_and_name,
    pipeline_set_actions,
]
```

8. We can use the list of aggregation stages to retrieve results and create the view in the `__main__` section of the script:

```
if __name__ == "__main__":
    client["beat_streaming"].drop_collection(
        "users_data_view"
    )

    client["beat_streaming"].create_collection(
        "users_data_view",
        viewOn="users",
        pipeline=pipeline,
    )
```

Say you run the script, from the terminal, for example, as follows:

```
$ python create_aggregation_and_user_data_view.py
```

The `users_data_view` view will be created in our `beat_streaming` database.

9. Once we have the view, we can create a dedicated endpoint to expose this view to a third-party customer without exposing any sensible data. We can create our endpoint in a separate module for clarity. In the app folder, let's create the `third_party_endpoint.py` module. In the module, let's create the module router as follows:

```
from fastapi import APIRouter, Depends

from app.database import mongo_database

router = APIRouter(
    prefix="/thirdparty",
    tags=["third party"],
)
```

10. Then, we can define the endpoint:

```
@router.get("/users/actions")
async def get_users_with_actions(
    db=Depends(mongo_database),
):
    users = [
        user
        async for user in db.users_data_view.find(
            {}, {"_id": 0}
        )
    ]
    return users
```

11. Once the endpoint function has been created, let's include the new router in the `FastAPI` object in the `main.py` module:

```
from app import third_party_endpoint
## rest of the main.py code
app = FastAPI(lifespan=lifespan)
app.include_router(third_party_endpoint.router)
## rest of the main.py code
```

The endpoint is now implemented in our API. Let's start the server by running the following command:

```
$ uvicorn app.main:app
```

At `http://localhost:8000/docs`, you can check that the newly created endpoint is present and call it to retrieve all the users from the created view without any sensible information.

You have just created an endpoint that securely exposes users data. An additional layer of security can be added by implementing **role-based access control** (**RBAC**) on the endpoint as we have done, for example, in *Chapter 4, Authentication and Authorization*, in the recipe *Setting up RBAC*.

There's more...

Additional layers are often added to secure your data's application, besides data masking. The most important ones are as follows:

- **Encryption at rest**
- **Encryption in transit**
- **RBAC**

The three services are provided as ready-to-use solutions in enterprise versions of MongoDB. The choice of using it or not is at the discretion of software architects.

Encryption at rest involves encrypting the data stored in your MongoDB database to prevent unauthorized access to sensitive information. The enterprise version of MongoDB provides built-in encryption capabilities through the use of a dedicated storage engine. By enabling encryption at rest, you can ensure that your data is encrypted on disk, making it unreadable to anyone without the proper encryption keys.

Encryption in transit ensures that data transmitted between your application and the MongoDB server is encrypted to prevent eavesdropping and tampering. MongoDB supports encryption in transit using **Transport Layer Security** (**TLS**), which encrypts data sent over the network between your application and the MongoDB server.

RBAC is essential for restricting access to sensitive data in MongoDB databases. MongoDB provides robust authentication and authorization mechanisms to control access to databases, collections, and documents. You can create user accounts with different roles and privileges to ensure that only authorized users can access and manipulate sensitive data.

MongoDB supports RBAC, allowing you to assign specific roles to users based on their responsibilities and restrict access to sensitive data accordingly.

See also

In the recipe, we had a quick look at how to create aggregations and views in MongoDB. Feel free to look into this more on the official documentation pages:

- *MongoDB Aggregations Quickstart*: `https://www.mongodb.com/developer/languages/python/python-quickstart-aggregation/`

- *MongoDB Views Documentation*: `https://www.mongodb.com/docs/manual/core/views/`

A good example of pushing data masking forward through database aggregations in MongoDB can be found at this link:

- *MongoDB Data Masking Example*: `https://github.com/pkdone/mongo-data-masking?tab=readme-ov-file`

You can see more about the commands of the aggregation framework on the official documentation page:

- *Aggregation Stage*: `https://www.mongodb.com/docs/manual/reference/operator/aggregation-pipeline/`

Also, a comprehensive book on MongoDB aggregations, free to consult, is available at this link:

- *Practical MongoDB Aggregation Book*: `https://www.practical-mongodb-aggregations.com`

Integrating FastAPI with Elasticsearch

Elasticsearch is a powerful search engine that provides fast and efficient full-text search, real-time analytics, and more. By integrating Elasticsearch with FastAPI, you can enable advanced search functionality, including keyword search, filtering, and aggregation. We'll walk through the process of integrating Elasticsearch, indexing data, executing search queries, and handling search results within a FastAPI application.

In this recipe, we will create a specific endpoint for our streaming platform to enable analytics and enhance search capabilities in your web applications. Specifically, we will retrieve the top ten artists based on views from a specified country.

By the end of this recipe, you'll be equipped with the knowledge and tools to leverage Elasticsearch for robust search functionality in your FastAPI projects.

Getting ready

To follow along with the recipe, you need a running application or to keep on working on our streaming platform.

Furthermore, you need an Elasticsearch instance running and reachable at this address: `http://localhost:9200`.

You can also install Elasticsearch on your machine by following the official guide: `https://www.elastic.co/guide/en/elasticsearch/reference/current/install-elasticsearch.html`.

Then, if you haven't installed the packages with `requirements.txt`, you need to install the Elasticsearch Python client with the `aiohttp` package in your environment. You can do this with `pip` from the command line:

```
$ pip install "elasticsearch>=8,<9" aiohttp
```

A basic knowledge of **Domain Specific Language** (**DSL**) in Elasticsearch can be beneficial to get a deeper understanding of the queries we are going to implement.

Have a look at the official documentation at this link: `https://www.elastic.co/guide/en/elasticsearch/reference/current/query-dsl.html`.

Once you have Elasticsearch installed and running, we can proceed to integrate it into our application.

How to do it...

We break down the process into the following steps:

1. Set up Elasticsearch in our FastAPI application to allow our API to communicate with the Elasticsearch instance.
2. Create an Elasticsearch index so that our songs can be indexed and queried by Elasticsearch.
3. Build the query to query our songs index.
4. Create the FastAPI endpoint to expose our analytics endpoint to the API users.

Let's look at each of these steps in detail.

Set up Elasticsearch in our FastAPI application

To interact with the Elasticsearch server, we need to define the client in our Python code. In the `db_connection.py` module, where we already define parameters for MongoDB, let's define the Elasticsearch asynchronous client:

```
from elasticsearch import AsyncElasticsearch,

es_client = AsyncElasticsearch(
    "localhost:9200"
)
```

We can create a function to check the connection with Elasticsearch in the same module:

```
from elasticsearch import (
    TransportError,
)

async def ping_elasticsearch_server():
    try:
        await es_client.info()
        logger.info(
            "Elasticsearch connection successful"
        )
    except TransportError as e:
        logger.error(
            f"Elasticsearch connection failed: {e}"
        )
        raise e
```

The function will ping the Elasticsearch server and propagate an error if the ping fails.

Then, we can call the function at the FastAPI server startup in the `lifetime` context manager in the `main.py` module:

```
@asynccontextmanager
async def lifespan(app: FastAPI):
    await ping_mongo_db_server(),
    await ping_elasticsearch_server()
# rest of the code
```

This will ensure that the application checks the connection with the Elasticsearch server at the startup, and it will propagate an error if the Elasticsearch server does not respond.

Create an Elasticsearch index

First of all, we should start by filling our Elasticsearch instance with a collection of song documents. In Elasticsearch, a collection is referred to as an *index*.

The song document should contain an additional field that tracks information about the views per country. For example, a new document song will look like the following:

```
{
    "title": "Song Title",
    "artist": "Singer Name",
    "album": {
```

```
        "title": "Album Title",
        "release_year": 2012,
        },
        "genre": "rock pop",
        "views_per_country": {
        "India": 50_000_000,
        "UK": 35_000_150_000,
        "Mexico": 60_000_000,
        "Spain": 40_000_000,
        },
    }
```

You can find a list of sampling songs in the file `songs_info.py` in the project GitHub repository. If you use the file, you can also define a function to fill in the index as:

```python
from app.db_connection import es_client

async def fill_elastichsearch():
    for song in songs_info:
        await es_client.index(
            index="songs_index", body=song
        )
    await es_client.close()
```

To group our songs based on the country's views, we will need to fetch data based on the `views_per_country` field, and for the top ten artists, we will group based on the `artist` field.

This information should be provided to the indexing process so that Elasticsearch understands how to index documents within the index for running queries.

In a new module called `fill_elasticsearch_index.py`, we can store this information in a python dictionary:

```python
mapping = {
    "mappings": {
        "properties": {
            "artist": {"type": "keyword"},
            "views_per_country": {
                "type": "object",
                "dynamic": True,
            },
        }
```

```
        }
    }
```

The `mapping` object will be passed as an argument to the Elasticsearch client when creating the index. We can define a function to create our `songs_index`:

```
from app.db_connection import es_client

async def create_index():
    await es_client.options(
        ignore_status=[400, 404]
    ).indices.create(
        index="songs_index",
        body=mapping,
    )
    await es_client.close()
```

You can run the function in into a grouping `main()` one, and use the `__main__` section of the module to run as follows:

```
async def main():
    await create_index()
    await fill_elastichsearch() # only if you use it

if __name__ == "__main__":
    import asyncio
    asyncio.run(create_index())
```

You can then run the script from the terminal:

```
$ python fill_elasticsearch_index.py
```

Now that the index is created, we just have to add the songs to the index. You can do this by creating a separate script or by running `fill_elasticsearch_index.py`, which is provided in the GitHub repository.

We have just set up our index filled with documents on our Elasticsearch index. Let's see how to build the query.

Build the query

We will build a function to return the query based on the specified country.

We can do it in a separate module in the `app` folder called `es_queries.py`. The query should fetch all the documents containing the `views_per_country` map index for the country and sort the results in descending order:

```
def top_ten_songs_query(country) -> dict:
    views_field = f"views_per_country.{country}"
    query = {
        "bool": {
            "must": {"match_all": {}},
            "filter": [
                {"exists": {"field": views_field}}
            ],
        }
    }
    sort = {views_field: {"order": "desc"}}
```

Then, we filter the fields that we want in the response as follows:

```
source = [
    "title",
    views_field,
    "album.title",
    "artist",
]
```

Finally, we return the query in the form of a dictionary by specifying the size of the list we will expect:

```
return {
    "index": "songs_index",
    "query": query,
    "size": 10,
    "sort": sort,
    "source": source,
}
```

We now have the function that will construct the query to retrieve the top ten artists for a specified country, and we will utilize it in our endpoint.

Create the FastAPI endpoint

Once we have set up the Elasticsearch connection and formulated the query, creating the endpoint is a straightforward process. Let's define it in a new module called `main_search.py` under the `app` folder. Let's start by defining the router:

```
from fastapi import APIRouter
router = APIRouter(prefix="/search", tags=["search"])
```

Then, the endpoint will be:

```
from fastapi import Depends, HTTPException
from app.db_connection import es_client

def get_elasticsearch_client():
    return es_client

@router.get("/top/ten/artists/{country}")
async def top_ten_artist_by_country(
    country: str,
    es_client=Depends(get_elasticsearch_client),
):
    try:
        response = await es_client.search(
         *top_ten_artists_query(country)
        )
    except BadRequestError as e:
        logger.error(e)

        raise HTTPException(
            status_code=400,
            detail="Invalid country",
        )

    return [
        {
            "artist": record.get("key"),
            "views": record.get("views", {}).get(
                "value"
            ),
        }
        for record in response["aggregations"][
            "top_ten_artists"
```

```
        ]["buckets"]
    ]
```

The result of the query is further adjusted before being returned to extract only the values we are interested in, namely the artist and views.

The last step is to include the router in our `FastAPI` object to include the endpoint.

In the `main.py` module, we can add the router as follows:

```
import main_search

## existing code in main.py
app = FastAPI(lifespan=lifespan)
app.include_router(third_party_endpoint.router)
app.include_router(main_search.router)

## rest of the code
```

Now, if you spin up the server with the `uvicorn app.main:app` command and go to the interactive documentation at `http://localhost:8000/docs`, you will see the newly created endpoint to retrieve the top ten artists in a country based on the views of the songs.

You have just created a FastAPI endpoint that interacts with an Elasticsearch instance. Feel free to create new endpoints on your own. For example, you can create an endpoint to return the top ten songs for a country.

See also

Since we have used the Elasticsearch Python client, feel free to dig more into the official documentation pages:

- *Elasticsearch Python Client*: `https://www.elastic.co/guide/en/elasticsearch/client/python-api/current/index.html`

- *Using Asyncio with Elasticsearch*: `https://elasticsearch-py.readthedocs.io/en/7.x/async.html`

To learn more about Elasticsearch indexes, have a look at the Elasticsearch documentation:

- *Index API*: `https://www.elastic.co/guide/en/elasticsearch/reference/current/docs-index_.html`

You can find a guide to mapping at this link:

- *Mapping*: https://www.elastic.co/guide/en/elasticsearch/reference/current/mapping.html

Finally, you can dig into the search query language at the following link:

- *Query DSL*: https://www.elastic.co/guide/en/elasticsearch/reference/current/query-dsl.html

Using Redis for caching in FastAPI

Redis is an in-memory data store that can be used as a cache to improve the performance and scalability of FastAPI applications. By caching frequently accessed data in Redis, you can reduce the load on your database and speed up response times for your API endpoints.

In this recipe, we'll explore how to integrate Redis caching into our streaming platform application and we will cache an endpoint as an example.

Getting ready

To follow along with the recipe you need a running Redis instance reachable at the http://localhost:6379 address.

Depending on your machine and your preference, you have several ways to install it and run it. Have a look at the Redis documentation to see how to do it for your operating system: https://redis.io/docs/install/install-redis/.

In addition, you need a FastAPI application with an endpoint that is time consuming.

Alternatively, if you follow the streaming platform, make sure that you have created the top ten artists endpoint from the previous recipe, *Integrating FastAPI with Elasticsearch*.

You will also need the Redis client for Python in your environment. If you haven't installed the packages with requirements.txt, you do it by running the following command:

```
$ pip install redis
```

Once the installation is complete, we can proceed with the recipe.

How to do it...

Once Redis is running and reachable at `localhost:6379`, we can integrate the Redis client into our code:

1. In the `db_connection.py` module, where we already defined the clients for Mongo and Elasticsearch, let's add the client for Redis:

   ```
   from redis import asyncio as aioredis

   redis_client = aioredis.from_url("redis://localhost")
   ```

2. Similarly to the other databases, we can create a function that pings the Redis server at the application's startup. The function can be defined as follows:

   ```
   async def ping_redis_server():
       try:
           await redis_client.ping()
           logger.info("Connected to Redis")
       except Exception as e:
           logger.error(
               f"Error connecting to Redis: {e}"
           )
           raise e
   ```

3. Then, include it in the `lifespan` context manager in `main.py`:

   ```
   @asynccontextmanager
   async def lifespan(app: FastAPI):
       await ping_mongo_db_server(),
       await ping_elasticsearch_server(),
       await ping_redis_server(),

       yield
   ```

   Now, we can use the `redis_client` object to cache our endpoints. We will cache the GET `/search/top/ten/artists` endpoint used to query Elasticsearch.

4. In `main_search.py`, we can define a function to retrieve the Redis client as a dependency:

   ```
   def get_redis_client():
       return redis_client
   ```

5.   Then, you can modify the endpoint as follows:

```
@router.get("/top/ten/artists/{country}")
async def top_ten_artist_by_country(
    country: str,
    es_client=Depends(get_elasticsearch_client),
    redis_client=Depends(get_redis_client),
):
```

6.   At the beginning of the function, we retrieve the key to store the value and check whether the value is already stored in Redis:

```
cache_key = f"top_ten_artists_{country}"

cached_data = await redis_client.get(cache_key)
if cached_data:
    logger.info(
        f"Returning cached data for {country}"
    )
    return json.loads(cached_data)
```

7.   Then, when we see that the data is not present, we continue by getting the data from Elasticsearch:

```
try:
    response = await es_client.search(
        *top_ten_artists_query(country)
    )
except BadRequestError as e:
    logger.error(e)

    raise HTTPException(
        status_code=400,
        detail="Invalid country",
    )

artists = [
    {
        "artist": record.get("key"),
        "views": record.get("views", {}).get(
            "value"
        ),
    }
    for record in response["aggregations"][
```

```
                    "top_ten_artists"
              ] ["buckets"]
        ]
```

8. Once we retrieve the list, we store it in Redis so we can retrieve it at the following call:

```
await redis_client.set(
    cache_key, json.dumps(artists), ex=3600
)

return artists
```

9. We specified an expiring time, which is the time the record will stay in Redis in seconds. After that time, the record won't be available anymore and the artists list will be recalled from Elasticsearch.

Now, if you run the server with the `uvicorn app.main:app` command and try to call the endpoint for Italy, you will notice that the response time for the second call will be much less.

You have just implemented a cache for one of the endpoints of our application with Redis. With the same strategy, feel free to cache all the other endpoints.

There's more...

At the time of writing, there is a promising library, `fastapi-cache`, which makes caching in FastAPI very easy. Check the GitHub repository: `https://github.com/long2ice/fastapi-cache`.

The library supports several caching databases, including Redis and in-memory caching. With simple endpoint decorators, you can specify caching parameters such as time to live, encoder, and cache response header.

See also

Redis client for Python supports more advanced functionalities. Feel free to explore its potential in the official documentation:

- *Redis Python Client*: `https://redis.io/docs/connect/clients/python/`
- *Redis Python Asynchronous Client*: `https://redis-py.readthedocs.io/en/stable/examples/asyncio_examples.html`

8

Advanced Features and Best Practices

Welcome to *Chapter 8*, where we explore advanced techniques and best practices to optimize the functionality, performance, and scalability of FastAPI applications.

In this chapter, by building a trip agency platform, you'll delve into essential topics such as dependency injection, custom middleware, internationalization, performance optimization, rate limiting, and background task execution. By mastering these advanced features, you'll be equipped to build robust, efficient, and high-performing APIs with FastAPI.

By the end of this chapter, you'll have a comprehensive understanding of the advanced FastAPI features and best practices, empowering you to build efficient, scalable, and secure APIs that meet the demands of modern web applications. Let's dive in and explore these advanced techniques to elevate your FastAPI development skills.

In this chapter, we're going to cover the following recipes:

- Implementing dependency injection
- Creating custom middleware
- Internationalization and localization
- Optimizing application performance
- Implementing rate limiting
- Implementing background tasks

Technical requirements

To be able to follow the recipes in this chapter, you must have a good grasp of the following essentials:

- **Python**: You should have a good understanding of Python version 3.7 or higher. You should know how annotation works and about basic class inheritance.
- `fastapi` **and** `asyncio` **libraries**: If you are not following the book chapter by chapter, make sure you know how to build a simple FastAPI endpoint and understand `async/await` syntax.

The code used in the chapter is hosted on GitHub at `https://github.com/PacktPublishing/FastAPI-Cookbook/tree/main/Chapter08`.

To manage dependencies more efficiently and keep your project isolated, consider creating a virtual environment within the `project` root folder. You can easily install all the dependencies simultaneously by using the `requirements.txt` file provided on the GitHub repository in the `project` folder:

```
$ pip install -r requirements.txt
```

You can then start with the first recipe and efficiently implement dependency injection in your FastAPI application.

Implementing dependency injection

Dependency injection is a powerful design pattern used in software development to manage dependencies between components. In the context of FastAPI, dependency injection allows you to efficiently manage and inject dependencies, such as database connections, authentication services, and configuration settings, into your application's endpoints and middleware. Although we have already used dependency injection in previous recipes, such as *Setting up SQL databases* in *Chapter 2, Working with Data*, or *Setting up user registration* in *Chapter 4, Authentication and Authorization*, this recipe will show you how to implement dependency injections in FastAPI and how to tackle trickier use cases with nested dependency injections.

Getting ready

To follow along with the recipe, you only need to have Python installed with the `fastapi` and `uvicorn` packages installed in your environment, as well as `pytest`. If you haven't installed the packages with the `requirements.txt` file provided in the GitHub repository, you can install them with `pip` from the command line:

```
$ pip install fastapi uvicorn pytest
```

Also, it would be beneficial to know already how to create a simple server in FastAPI. You can refer to the *Creating a new FastAPI project* recipe in *Chapter 1, First Steps with FastAPI*, for more details.

How to do it...

Let's start by creating the project root folder called `trip_platform` containing the `app` folder.
Then let's continue the recipe through the following steps.

1. In the `app` folder, create the `main.py` module that will contain the server as:

    ```python
    from fastapi import FastAPI


    app = FastAPI()
    ```

    We will write the dependencies in a separate module called `dependencies.py` inside the
    `app` folder.

2. Let's imagine we need to create an endpoint to retrieve all the trips between a start date and an
    end date. We need to handle two parameters, the start date and end date, and check that the
    start date is earlier than the end date. Both parameters can be optional; if the start date is not
    provided, it defaults to the current day.

    In a dedicated module, `dependencies.py` in the `app` folder, let's define the condition
    function, which checks that the start date is earlier than the end date:

    ```python
    from fastapi import HTTPException


    def check_start_end_condition(start: date, end: date):
        if end and end < start:
            raise HTTPException(
                status_code=400,
                detail=(
                    "End date must be "
                    "greater than start date"
                ),
            )
    ```

3. We use the `check_start_end_condition` function to define the `dependable` function
    – namely, the function that will be used as a dependency – as follows:

    ```python
    from datetime import date, timedelta
    from fastapi import Query


    def time_range(
        start: date | None = Query(
            default=date.today(),
            description=(
                "If not provided the current date is used"
    ```

```
        ),
            example=date.today().isoformat(),
        ),
    end: date | None = Query(
        None,
        example=date.today() + timedelta(days=7),
    ),
) -> Tuple[date, date | None]:
    check_start_end_condition(start, end)
    return start, end
```

The `Query` object is used to manage metadata of the query parameters, such as the default value, description, and example used when generating the documentation.

4.   We can use the dependable `time_range` function to create the endpoint in the `main.py` module. To specify that it is a dependency, we use the `Depends` object like this:

```
from fastapi import Depends

@app.get("/v1/trips")
def get_tours(
    time_range = Depends(time_range),
):
    start, end = time_range
    message = f"Request trips from {start}"
    if end:
        return f"{message} to {end}"
    return message
```

You can alternatively use the `Annotated` class from the `typing` package to define the dependency as follows:

```
from typing import Annotated
from fastapi import Depends

@app.get("/v1/trips")
def get_tours(
    time_range: Annotated[time_range, Depends()]
):
```

Important note

The use of Annotated in FastAPI is currently evolving to avoid duplicates and improve readability; take a look at the dedicated documentation section: https://fastapi.tiangolo.com/tutorial/dependencies/#share-annotated-dependencies.

For the rest of the chapter, we will use the latest Annotated convention.

Now, if you spin up the server by running uvicorn app.main:app on the terminal, you will find the endpoint in the interactive documentation at http://localhost:8000/docs. You will see that you just created the endpoint with the parameters correctly documented. The database logic is replaced by a string construction returning a significant message in the example.

You have just implemented a dependency injection strategy to define query parameters for the endpoint. You can use the same strategy to write path or body parameters to write modular and readable code.

One of the advantages of using dependency injection is to logically separate pieces of code that can be replaced by something else, like in testing. Let's have a look at how to do it.

Overriding dependency injections in tests

Let's create a test for the GET /v1/trips endpoint. If you don't have pytest in the environment, install it with pip install pytest. Then, under the project root folder, create the pytest.ini file containing pythonpath for pytest, as follows:

```
[pytest]
pythonpath=.
```

The test will be in the test module, test_main.py, under the tests folder. Let's write a unit test by overriding the client's dependency:

```
from datetime import date

from fastapi.testclient import TestClient

from app.dependencies import time_range
from app.main import app


def test_get_v1_trips_endpoint():
    client = TestClient(app)
    app.dependency_overrides[time_range] = lambda: (
        date.fromisoformat("2024-02-01"),
        None,
    )
    response = client.get("/v1/trips")
```

```
assert (
    response.json()
    == "Request trips from 2024-02-01"
)
```

By overriding the `time_range` dependency, we won't need to pass the parameters when calling the endpoint, and the response will depend on the lambda function defined.

Then, you can run the test from the command line:

```
$ pytest tests
```

This technique is very useful when writing tests that should not interfere with a production database. Also, an eventual heavy computation logic can be mocked if it is not in the test's interest.

The use of dependency injection can significantly improve test quality by enabling modularity.

How it works...

The `Depends` object and dependency injection leverage Python's powerful function annotations and type hinting features.

When you define a dependency function and annotate it with `Depends`, FastAPI interprets it as a dependency that needs to be resolved before executing the endpoint function. When a request is made to an endpoint that depends on one or more dependencies, FastAPI introspects the endpoint function signature, identifies the dependencies, and resolves them by invoking the corresponding dependency functions in the correct order.

FastAPI uses Python's type hinting mechanism to determine the type of each dependency parameter and automatically injects the resolved dependency into the endpoint function. This process ensures that the required data or services are available to the endpoint function at runtime, enabling seamless integration of external services, database connections, authentication mechanisms, and other dependencies into FastAPI applications.

Overall, the `Depends` class and dependency injection in FastAPI provide a clean and efficient way to manage dependencies and promote modular, maintainable code architecture. One of the advantages is that they can be overwritten in testing to be easily mocked or replaced.

There's more...

We can push things further by leveraging sub-dependencies.

Let's create an endpoint that returns the trips for one of the three categories (cruises, city breaks, and resort stays) and, simultaneously, checks the coupon validity for the category.

In the `dependencies.py` module, let's create the `dependable` function for the category.

Imagine we can group our trips into three categories – cruises, city breaks, and resort stays. We need to add a parameter to retrieve trips only for a specific category. We will need a `dependable` function, as follows:

```
def select_category(
    category: Annotated[
        str,
        Path(
            description=(
                "Kind of travel "
                "you are interested in"
            ),
            enum=[
                "Cruises",
                "City Breaks",
                "Resort Stay",
            ],
        ),
    ],
) -> str:
    return category
```

Now, let's imagine we need to validate a coupon for a discount.

The `dependable` function will be used as a dependency for another `dependable` function that will check the coupon. Let's define it, as follows:

```
def check_coupon_validity(
    category: Annotated[select_category, Depends()],
    code: str | None = Query(
        None, description="Coupon code"
    ),
) -> bool:
    coupon_dict = {
        "cruises": "CRUISE10",
        "city-breaks": "CITYBREAK15",
        "resort-stays": "RESORT20",
    }
    if (
        code is not None
        and coupon_dict.get(category, ...) == code
    ):
```

```
        return True
    return False
```

In the `main.py` module, let's define a new endpoint, `GET /v2/trips/{category}`, that returns the trips for the specified category:

```
@app.get("/v2/trips/{category}")
def get_trips_by_category(
    category: Annotated[select_category, Depends()],
    discount_applicable: Annotated[
        bool, Depends(check_coupon_validity)
    ],
):
    category = category.replace("-", " ").title()
    message = f"You requested {category} trips."

    if discount_applicable:
        message += (
            "\n. The coupon code is valid! "
            "You will get a discount!"
        )
    return message
```

If you run the server with the `uvicorn app.main:app` command and open the interactive documentation at `http://localhost:8000/docs`, you will see the new endpoint. The accepted parameters, `category` and `code`, both come from the dependencies, and the `category` parameter is not repeated within the code.

> **Important note**
> You can use both `def` and `async def` keywords to declare dependencies, whether they are synchronous or asynchronous functions respectively. FastAPI will handle them automatically.

You have just created an endpoint that uses nested dependencies. By using nested dependencies and sub-dependencies, you will be able to write clear and modular code that is easier to read and maintain.

> **Exercise**
> In FastAPI, dependencies can be also created as a class. Check out the documentation at `https://fastapi.tiangolo.com/tutorial/dependencies/classes-as-dependencies/#classes-as-dependencies`, and create a new endpoint that uses all the parameters we defined in the recipe (`time_range`, `category`, and `code`).
> Group all the parameters into a class, and define and use it as a dependency for the endpoint.

See also

We have used `Query` and `Path` descriptor objects to set metadata and documentation-related data for `query` and `path` parameters, respectively. You can discover more about their potential at these documentation links:

- *Query Parameters and String Validations*: `https://fastapi.tiangolo.com/tutorial/query-params-str-validations/`

- *Path Parameters and Numeric Validations*: `https://fastapi.tiangolo.com/tutorial/path-params-numeric-validations/`

For dependency injections in FastAPI, you can find extensive documentation covering all the possible usages, explaining the potential of this powerful feature:

- *Dependencies*: `https://fastapi.tiangolo.com/tutorial/dependencies/`

- *Advanced Dependencies*: `https://fastapi.tiangolo.com/advanced/advanced-dependencies/`

- *Testing Dependencies with Overrides*: `https://fastapi.tiangolo.com/advanced/testing-dependencies/`

Creating custom middleware

Middleware is an API component that allows you to intercept and modify incoming requests and outgoing responses, making it a powerful tool for implementing cross-cutting concerns such as authentication, logging, and error handling.

In this recipe, we'll explore how to develop custom middleware to process requests and responses in FastAPI applications and retrieve information on the client.

Getting ready...

All you need is to have a running FastAPI application. The recipe will use our trip platform defined in the previous recipe, *Implementing dependency injection*. However, middleware works for a generic running application.

How to do it...

We will show you how to create a custom middleware object class that we will use in our application through the following steps.

1. Let's create a dedicated module in the app folder called `middleware.py`.

We want the middleware to intercept the request and print the host client and the method on the output terminal. In a real application scenario, this information can be stored in a database for analytics or used for security inspection purposes.

Let's use the same `uvicorn` logger used by FastAPI by default:

```
import logging

logger = logging.getLogger("uvicorn.error")
```

2. Then, let's create our `ClientInfoMiddleware` class, as follows:

```
from fastapi import Request
from starlette.middleware.base import BaseHTTPMiddleware

class ClientInfoMiddleware(BaseHTTPMiddleware):
    async def dispatch(
        self, request: Request, call_next
    ):
        host_client = request.client.host
        requested_path = request.url.path
        method = request.method

        logger.info(
            f"host client {host_client} "
            f"requested {method} {requested_path} "
            "endpoint"
        )

        return await call_next(request)
```

3. Then, we need to add our middleware to the FastAPI server in `main.py`. After defining the app server, we can add the middleware with the `add_middleware` method:

```
# main.py import modules
from app.middleware import ClientInfoMiddleware

app = FastAPI()

app.add_middleware(ClientInfoMiddleware)

# rest of the code
```

Now, spin up the server with the `uvicorn app.main:app` command, and try to connect to a subpath of `http://localhost:8000/v1/trips`. You don't even need to call an existing endpoint. You will see the log messages in the application output terminal:

```
INFO:host client 127.0.0.1 requested GET /v1/trips endpoint
```

You have just implemented a basic custom middleware to retrieve information about the client. You can increase the complexity by adding more operations, such as redirecting requests based on the IP and integrating IP blocking or filtering.

How it works...

FastAPI uses the `BasicHTTPMiddleware` class from the `Starlette` library. The strategy shown in the recipe creates a class derived from `BasicHTTPMiddleware`, with a specific `dispatch` method that implements the interception operation.

To create a middleware in FastAPI, you can add a decorator from the FastAPI class methods to a simple function. However, it is recommended to create a class, as it allows for better modularity and organization of the code. By creating a class, you can eventually create your collection module of middleware.

See also

You can have a look at how to create a custom middleware on the official documentation page at the following link:

- *FastAPI Middleware documentation*: `https://fastapi.tiangolo.com/tutorial/middleware/`

An interesting discussion on how to create middleware classes in FastAPI can be found on the **Stack Overflow** website:

- *Create FastAPI Custom Middleware Class Discussion*: `https://stackoverflow.com/questions/71525132/how-to-write-a-custom-fastapi-middleware-class`

Internationalization and localization

Internationalization (i18n) and **localization (l10n)** are fundamental concepts in software development that enable applications to be adapted for different languages, regions, and cultures.

i18n refers to the process of designing and developing software or products that can be adapted to different languages and cultures. This process mainly involves providing content in a specific language. Conversely, **l10n** involves adapting a product or content for a specific locale or market, such as currency or a unit of measure.

The **Accept-Language request HTTP header** is the most commonly used method to inform a server about a user's location. This is widely used by modern browsers. This recipe will guide you on how to utilize the `Accept-Language` header to implement i18n and l10n in our trip platform. This will enable our platform to provide targeted content to the client.

Getting ready

It would be beneficial to have some knowledge of the `Accept-Language` header; take a look at this interesting article from Mozilla's documentation: `https://developer.mozilla.org/en-US/docs/Web/HTTP/Headers/Accept-Language`.

You need to have a running FastAPI application to follow the recipe. You can follow along with the trip platform application used throughout the chapter.

We will make use of dependency injection, so it will be beneficial to complete the *Implementing dependency injection* recipe from this chapter.

Also, we will use the `babel` package to resolve language code references, so if haven't installed the packages with the `requirements.txt` file, make sure to have `babel` in your environment by running the following:

```
$ pip install babel
```

Once the installation is completed, you have all you need to start the recipe.

How to do it...

To begin with, we must determine which regions and languages we wish to cater to. For this example, we will focus on two – **American English (en_US)** and **French from France (fr_FR)**. All content pertaining to language will be in one of these two languages.

It is necessary to manage the `Accept-Language` header on the host client side, which is a list of languages with a preference weight parameter.

Examples of the header are as follows:

```
Accept-Language: en
Accept-Language: en, fr
Accept-Language: en-US
Accept-Language: en-US;q=0.8, fr;q=0.5
Accept-Language: en, *
Accept-Language: en-US, en-GB
Accept-Language: zh-Hans-CN
```

We need a function that takes as an argument the header and the list of available languages in our app, returning the most appropriate one Let's implement it by applying the following steps.

1. Create a dedicated module, `internationalization.py`, under the app folder.

    First, we store the supported languages in a variable, as follows:

    ```
    SUPPORTED_LOCALES = [
        "en_US",
        "fr_FR",
    ]
    ```

2. Then, we start defining the `resolve_accept_lanugage` function, as follows:

    ```
    from babel import Locale, negotiate_locale

    def resolve_accept_language(
        accept_language: str = Header("en-US"),
    ) -> Locale:
    ```

3. Within the function, we parse the string into a list:

    ```
    client_locales = []
    for language_q in accept_language.split(","):
        if ";q=" in language_q:
            language, q = language_q.split(";q=")
        else:
            language, q = (language_q, float("inf"))
        try:
            Locale.parse(language, sep="-")
            client_locales.append(
                (language, float(q))
            )
        except ValueError:
            continue
    ```

4. We then sort the string according to the preference q parameter:

    ```
    client_locales.sort(
        key=lambda x: x[1], reverse=True
    )

    locales = [locale for locale, _ in client_locales]
    ```

5. Then, we use `negotiate_locale` from the `babel` package to get the most suited language:

    ```
    locale = negotiate_locale(
        [str(locale) for locale in locales],
        SUPPORTED_LOCALES,
    )
    ```

6. If there is no match, we return en_US as default:

    ```
    if locale is None:
        locale = "en_US"

    return locale
    ```

The `resolve_accept_language` function will be used as a dependency for the endpoints that return content based on the language.

7. In the same `internationalization.py` module, let's create a GET /homepage endpoint that returns a welcome string, depending on the language. We will do it in a separate `APIRouter`, so the router will be as follows:

    ```
    from fastapi import APIRouter

    router = APIRouter(
        tags=["Localizad Content Endpoints"]
    )
    ```

The `tags` parameter specifies that the router's endpoint will be grouped separately in the interactive documentation under a specified tag name.

The GET /home endpoint will be as follows:

```
home_page_content = {
    "en_US": "Welcome to Trip Platform",
    "fr_FR": "Bienvenue sur Trip Platform",
}

@router.get("/homepage")
async def home(
    request: Request,
    language: Annotated[
        resolve_accept_language, Depends()
    ],
):
    return {"message": home_page_content[language]}
```

In the example, the content has been hardcoded as a `dict` object with language code as a dictionary key.

In a real-world scenario, the content should be stored in a database for each language.

Similarly, you define a localization strategy to retrieve the currency.

8. Let's create a GET `/show/currency` endpoint as an example that uses a dependency to retrieve the currency from the `Accept-Language` header. The dependency function can be defined as follows:

```python
async def get_currency(
    language: Annotated[
        resolve_accept_language, Depends()
    ],
):
    currencies = {
        "en_US": "USD",
        "fr_FR": "EUR",
    }

    return currencies[language]
```

The endpoint will then be as follows:

```python
from babel.numbers import get_currency_name

@router.get("/show/currency")
async def show_currency(
    currency: Annotated[get_currency, Depends()],
    language: Annotated[
        resolve_accept_language,
        Depends(use_cache=True)
    ],
):
    currency_name = get_currency_name(
        currency, locale=language
    )
    return {
        "currency": currency,
        "currency_name": currency_name,
    }
```

9. To use both endpoints, we will need to add the router to the FastAPI object in `main.py`:

```
from app import internationalization
# rest of the code


app.include_router(internationalization.router)
```

This is all you need to implement internationalization and localization. To test it, spin up the server from the command line by running:

```
$ uvicorn app.main:app
```

On the interactive documentation at `http:localhost:8000/docs`, you will find the `GET /homepage` and `GET /show/currency` endpoints. Both accept the `Accept-Language` header to provide the language choice; if you don't, it will get the default language from the browser. To test the implementation, try experimenting with different values for the header.

You have successfully implemented internationalization and localization from scratch for your API. Using the recipe provided, you have integrated i18n and l10n into your applications, making them easily understandable worldwide.

See also

You can find out more about the potential of `Babel` package on the official documentation page: `https://babel.pocoo.org/en/latest/`.

Optimizing application performance

Optimizing FastAPI applications is crucial for ensuring high performance and scalability, especially under heavy loads.

In this recipe, we'll see a technique to profile our FastAPI application and explore actionable strategies to optimize performances. By the end of the recipe, you will be able to detect code bottlenecks and optimize your application.

Getting ready

Before starting the recipe, make sure to have a FastAPI application running with some endpoints already set up. You can follow along with our trip platform application.

We will be using the `pyinstrument` package to set up a profiler for the application. If you haven't installed the packages with `requirements.txt`, you can install `pyinstrument` in your environment by running the following:

```
$ pip install pyinstrument
```

Also, it can be useful to have a look at the *Creating custom middleware* recipe from earlier in the chapter.

How to do it...

Let's implement the profiler in simple steps.

1. Under the app folder, create a `profiler.py` module as follows:

    ```
    from pyinstrument import Profiler


    profiler = Profiler(
        interval=0.001, async_mode="enabled"
    )
    ```

    The `async_mode="enabled"` parameter specifies that the profiler logs the time each time it encounters an `await` statement in the function being awaited, rather than observing other coroutines or the event loop. The `interval` specifies the time between two samples.

2. Before using the profiler, we should plan what we want to profile. Let's plan to profile only the code executed in the endpoints. To do this, we can create simple middleware in a separate module that starts and stops the profiler before and after each call, respectively. We can create the middleware in the same `profiler.py` module, as follows:

    ```
    from starlette.middleware.base import (
        BaseHTTPMiddleware
    )


    class ProfileEndpointsMiddleWare(
        BaseHTTPMiddleware
    ):
        async def dispatch(
            self, request: Request, call_next
        ):
            if not profiler.is_running:
                profiler.start()
            response = await call_next(request)
            if profiler.is_running:
                profiler.stop()
                profiler.write_html(
    ```

```
                    os.getcwd() + "/profiler.html"
            )
        profiler.start()
    return response
```

The profiler is initiated every time an endpoint is requested, and it is terminated after the request is complete. However, since the server operates asynchronously, there is a possibility that the profiler may already be running, due to another endpoint request. This can result in errors during the start and stop of the profiler. To prevent this, we verify before each request whether the profiler is not already running. After the request, we check whether the profiler is running before terminating it.

3. You can attach the profiler to the FastAPI server by adding the middleware in the `main.py` module, as we did in the *Creating custom middleware* recipe:

```
app.add_middleware(ProfileEndpointsMiddleWare)
```

To test the profiler, spin up the server by running `uvicorn app.main:app`. Once you start making some calls, you can do it from the interactive documentation at `http://localhost:8000/docs`. Then, a `profiler.html` file will be created. You can open the file with a simple browser and check the status of the code.

You have just integrated a profiler into your FastAPI application.

There's more...

Integrating a profiler is the first step that allows you to spot code bottlenecks and optimize the performance of your application.

Let's explore some techniques to optimize the performance of your FastAPI performances:

*   **Asynchronous programming**: Utilize asynchronous programming to handle concurrent requests efficiently. FastAPI is built on top of the `Starlette` library and supports asynchronous request handlers, using the `async` and `await` keywords. By leveraging asynchronous programming, you can maximize CPU and **input/output** (**I/O**) utilization, reducing response times and improving scalability.

*   **Scaling Uvicorn workers**: Increasing the number of Uvicorn workers distributes incoming requests across multiple processes. However, it might not be always the best solution. For purely I/O operations, asynchronous programming massively reduces CPU usage, and additional workers remain idle. Before adding additional workers, check the CPU usage of the main process.

*   **Caching**: Implement caching mechanisms to store and reuse frequently accessed data, reducing database queries and computation overhead. Use dedicated libraries l to integrate caching into your FastAPI applications.

Other techniques are related to external libraries or tools, and whatever strategy you use, make sure to properly validate it with proper profiling configuration.

Also, for high-traffic testing, take a look at the *Performance testing for high traffic applications* recipe in *Chapter 5, Testing and Debugging FastAPI Applications*.

> **Exercise**
>
> We learned how to configure middleware to profile applications; however, it is more common to create tests to profile specific use cases. We learned how to configure middleware to profile applications; however, it is more common to create test scripts to profile specific use cases. Try to create one by yourself that attaches the profiler to the server, runs the server, makes API calls that reproduce the use case, and finally, writes the profiler output. The solution is provided on the GitHub repository in the `profiling_application.py` file. You can find it at `https://github.com/PacktPublishing/FastAPI-Cookbook/blob/main/Chapter08/trip_platform/profiling_application.py`.

See also

You can discover more about the potential of **pyinstrument** profiler on the official documentation:

- *pyinstrument documentation*: `https://pyinstrument.readthedocs.io/en/latest/`

Also, you can find a different approach to profile FastAPI endpoints on the page:

- *pyinstrument – profiling FastAPI requests*: `https://pyinstrument.readthedocs.io/en/latest/guide.html#profile-a-web-request-in-fastapi`

Implementing rate limiting

Rate limiting is an essential technique used to control and manage the flow of traffic to web applications, ensuring optimal performance, resource utilization, and protection against abuse or overload. In this recipe, we'll explore how to implement rate limiting in FastAPI applications to safeguard against potential abuse, mitigate security risks, and optimize application responsiveness. By the end of this recipe, you'll have a solid understanding of how to leverage rate limiting to enhance the security, reliability, and scalability of your FastAPI applications, ensuring optimal performance under varying traffic conditions and usage patterns.

Getting ready

To follow the recipe, you need a running FastAPI application with some endpoints to use for rate limiting. To implement rate limiting, we will use the `slowapi` package; if you haven't installed the packages with the `requirements.txt` file provided in the GitHub repository, you can install `slowapi` in your environment with `pip` by running the following:

```
$ pip install slowapi
```

Once the installation is completed, you are ready to start the recipe.

How to do it...

We will start by applying a rate limiter to a single endpoint in simple steps.

1. Let's create the `rate_limiter.py` module under the `app` folder that contains our limiter object class defined as follows:

    ```
    from slowapi import Limiter
    from slowapi.util import get_remote_address

    limiter = Limiter(
        key_func=get_remote_address,
    )
    ```

    The limiter is designed to restrict the number of requests from a client based on their IP address. It is possible to create a function that can detect a user's credentials and limit their calls according to their specific user profile. However, for the purpose of this example, we will use the client's IP address to implement the limiter.

2. Now, we need to configure the FastAPI server to implement the limiter. In `main.py`, we have to add the following configuration:

    ```
    from slowapi import _rate_limit_exceeded_handler
    from slowapi.errors import RateLimitExceeded

    # rest of the code

    app.state.limiter = limiter
    app.add_exception_handler(
        RateLimitExceeded, _rate_limit_exceeded_handler
    )

    # rest of the code
    ```

3. Now, we will apply a rate limit of two requests per minute to the GET /homepage endpoint defined in the internalization.py module:

```
from fastapi import Request
from app.rate_limiter import limiter


@router.get("/homepage")
@limiter.limit("2/minute")
async def home(
    request: Request,
    language: Annotated[
        resolve_accept_language, Depends()
    ],
):
    return {"message": home_page_content[language]}
```

The rate limit is applied as a decorator. Also, the request parameter needs to be added to make the limiter work.

Now, spin up the server from the command line by running the following:

```
$ uvicorn app.main:app
```

Then, try to make three consecutive calls to http://localhost:8000/homepage; you will get the home page content, and by the third call, you will get a 429 response with the following content:

```
{
    "error": "Rate limit exceeded: 2 per 1 minute"
}
```

You've just added a limit rate to the GET /homepage endpoint. With the same strategy, you can add a specific rate limiter to each endpoint.

There's more...

You can do more by adding a global rate limit to the entire application, as follows.

In main.py, you need to add a dedicated middleware, as follows:

```
# rest of the code in main.py
from slowapi.middleware import SlowAPIMiddleware

# rest of the code

app.add_exception_handler(
```

```
        RateLimitExceeded, _rate_limit_exceeded_handler
    )
app.add_middleware(SlowAPIMiddleware)
```

Then, you simply need to specify the default limit in the `Limiter` object instantiation in the `rate_limiter.py` module:

```
limiter = Limiter(
    key_func=get_remote_address,
    default_limits=["5/minute"],
)
```

And that's it. Now, if you rerun the server and call any of the endpoints more than five times consecutively, you will get a `429` response.

You have successfully set up a global rate limiter for your FastAPI application.

See also

You can find more on **Slowapi** features such as shared limits, limiting policies, and more in the official documentation at this link:

- *SlowApi documentation*: `https://slowapi.readthedocs.io/en/latest/`

You can check out more on the syntax of rate limit notation in the **Limits** project documentation at this link:

- *Rate limit string notation*: `https://limits.readthedocs.io/en/stable/quickstart.html#rate-limit-string-notation`

Implementing background tasks

Background tasks are a useful feature that enables you to delegate resource-intensive operations to separate processes. With background tasks, your application can remain responsive and handle multiple requests simultaneously. They are particularly important for handling long-running processes without blocking the main request-response cycle. This improves the overall efficiency and scalability of your application. In this recipe, we will explore how you can execute background tasks in FastAPI applications.

Getting ready

To follow this recipe, all you need is a FastAPI application running with at least one endpoint to apply the background task. However, we will implement the background task into our trip platform into the GET /v2/trips/{category} endpoint, defined in the *Implementing dependency injection* recipe.

How to do it...

Let's imagine we want to store the message of the GET /v2/trips/{category} endpoint in an external database for analytics purposes. Let's do it in two simple steps.

1. First, we define a function that mocks the storing operation in a dedicated module, background_tasks.py, in the app folder. The function will look like the following:

```python
import asyncio
import logging

logger = logging.getLogger("uvicorn.error")

async def store_query_to_external_db(message: str):
    logger.info(f"Storing message '{message}'.")
    await asyncio.sleep(2)
    logger.info(f"Message '{message}' stored!")
```

The storing operation is mocked by an asyncio.sleep non-blocking operation. We have also added some log messages to keep track of the execution.

2. Now, we need to execute the store_query_to_external_db function as a background task of our endpoint. In main.py, let's modify the GET /v2/trips/cruises, as follows:

```python
from fastapi import BackgroundTasks

@app.get("/v2/trips/{category}")
def get_trips_by_category(
    background_tasks: BackgroundTasks,
    category: Annotated[select_category, Depends()],
    discount_applicable: Annotated[
        bool, Depends(check_coupon_validity)
    ],
):
    category = category.replace("-", " ").title()
    message = f"You requested {category} trips."

    if discount_applicable:
        message += (
```

```
                "\n. The coupon code is valid! "
                "You will get a discount!"
        )


    background_tasks.add_task(
        store_query_to_external_db, message
    )
    logger.info(
        "Query sent to background task, "
        "end of request."
    )
    return message
```

Now, if you spin up the server with `uvicorn app.main:app` and try to call the `GET /v2/trips/cruises` endpoint, you will see the logs from the `store_query_to_external_db` function on the terminal output:

```
INFO:   Query sent to background task, end of request.
INFO:   127.0.0.1:58544 - "GET /v2/trips/cruises
INFO:   Storing message 'You requested Cruises trips.'
INFO:   Message 'You requested Cruises trips.' Stored!
```

That is all you need to implement background tasks in FastAPI! However, if you have to perform extensive background computations, you might want to use dedicated tools to handle queued task execution. This would allow you to run the tasks in a separate process and avoid any performance issues that may arise from running them in the same process.

How it works...

When a request is made to the endpoint, the background task is enqueued to the `BackgroundTasks` object. All the tasks are passed to the event loop so that they can be executed concurrently, allowing for non-blocking I/O operations.

If you have a task that requires a lot of processing power and doesn't necessarily need to be completed by the same process, you might want to consider using larger tools such as Celery.

See also

You can find more on creating background tasks in FastAPI on the official documentation page at this link:

- *Background Tasks*: `https://fastapi.tiangolo.com/tutorial/background-tasks/`

Working with WebSocket

Real-time communication has become increasingly important in modern web applications, enabling interactive features such as chat, notifications, and live updates. In this chapter, we'll explore the exciting world of WebSockets and how to leverage them effectively in FastAPI applications. From setting up WebSocket connections to implementing advanced features such as chat functionality and error handling, this chapter provides a comprehensive guide to building responsive, real-time communication features. By the end of the chapter, you will have the skills to create WebSockets and facilitate real-time communication in FastAPI applications, enabling interactive functionalities and dynamic user experiences.

In this chapter, we're going to cover the following recipes:

- Setting up Websockets in FastAPI
- Sending and receiving messages over WebSockets
- Handling WebSocket connections and disconnections
- Handling WebSocket errors and exceptions
- Implementing chat functionality with WebSocket
- Optimizing WebSocket performance
- Securing WebSocket connections with OAuth2

Technical requirements

To follow along with the WebSockets recipes, make sure you have the following essentials in your setup:

- **Python**: Install a Python version higher than 3.9 in your environment.
- **FastAPI**: This should be installed with all the required dependencies. If you haven't done it in the previous chapters, you can simply do it from your terminal:

```
$ pip install fastapi[all]
```

The code used in the chapter is hosted on GitHub at `https://github.com/PacktPublishing/FastAPI-Cookbook/tree/main/Chapter09`.

It is recommended to set up a virtual environment for the project in the project root folder to efficiently manage dependencies and maintain project isolation.

Within your virtual environment, you can install all the dependencies at once by using the `requirements.txt` file provided in the GitHub repository in the project folder:

```
$ pip install -r requirements.txt
```

Since the interactive Swagger documentation is limited at the time of writing, basic mastering of **Postman** or any other testing API is beneficial to test our API.

Having basic knowledge of how **WebSockets** work can be beneficial, although it's not necessary since the recipes will guide you through.

For the *Implementing chat functionality with WebSockets* recipe, we will write some basic **HTML**, including some **Javascript** code.

Setting up WebSockets in FastAPI

WebSockets provide a powerful mechanism for establishing full-duplex communication channels between clients and servers, allowing real-time data exchange. In this recipe, you'll learn how to establish a connection with WebSocket functionality in your FastAPI applications to enable interactive and responsive communication.

Getting ready

Before diving into the recipe, ensure you have all the required packages in your environment. You can install them from the `requirements.txt` file provided in the GitHub repository or install it manually with `pip`:

```
$ pip install fastapi[all] websockets
```

Since the swagger documentation does not support WebSocket, we will use an external tool to test the WebSocket connection, such as Postman.

You can find instructions on how to install it on the website:
`https://www.postman.com/downloads/`.

The free community version will be enough to test the recipes.

How to do it...

Create the project root folder called `chat_platform`. we can create our `app` folder containing the `main.py` module. Let's build our simple application with a WebSocket endpoint as follows.

1. We can start by creating our server in the `main.py` module:

```
from fastapi import FastAPI


app = FastAPI()
```

2. Then we can create the WebSocket endpoint to connect the client to the chat room:

```
from fastapi import WebSocket


@app.websocket("/ws")
async def ws_endpoint(websocket: WebSocket):
    await websocket.accept()
    await websocket.send_text(
        "Welcome to the chat room!"
    )
    await websocket.close()
```

The endpoint establishes the connection with the client, sends a welcome message, and closes the connection. This is the most basic configuration of a WebSocket endpoint.

That's it. To test it, spin up the server from the command line:

```
$ uvicorn app.main:app
```

Then open Postman and create a new WebSocket request. Specify the server URL as `ws://localhost:8000/ws` and click on **Connect**.

In the **Response** panel, right below the URL form, you should see the list of events that happened during the connection. In particular, look for the message received by the server:

```
Welcome to the chat room! 12:37:19
```

That means that the WebSocket endpoint has been created and works properly.

How it works...

The `websocket` parameter in the WebSocket endpoint represents an individual WebSocket connection. By awaiting `websocket.accept()`, the server establishes the connection with the client (technically called an **HTTP handshake**). Then, `websocket.send_text()` sends a message to the client. Finally, `websocket.close()` closes the connection.

The three events are listed in the **Response** panel of Postman.

Although not very useful from a practical point of view, this configuration is the bare minimum setup for a WebSocket connection. In the next recipe, we will see how to exchange messages between the client and the server through a WebSocket endpoint.

See also

You can check how to create a WebSocket endpoint on the FastAPI official documentation page:

- *FastAPI WebSockets*: `https://fastapi.tiangolo.com/advanced/websockets/`

At the time of writing, the Swagger documentation does not support WebSocket endpoints. If you spin up the server and open Swagger at `http://localhost:8000/docs`, you won't see the endpoint we have just created. A discussion is ongoing on the FastAPI GitHub repository – you can follow it at the following URL:

- *FastAPI WebSocket Endpoints Documentation Discussion*: `https://github.com/tiangolo/fastapi/discussions/7713`

Sending and receiving messages over WebSockets

WebSocket connections enable bidirectional communication between clients and servers, allowing the real-time exchange of messages. This recipe will bring us one step closer to creating our chat application by showing how to enable the FastAPI application to receive messages over WebSockets and print them to the terminal output.

Getting ready

Before starting the recipe, make sure you know how to set up a **WebSocket** connection in **FastAPI**, as explained in the previous recipe. Also, you will need a tool to test WebSockets, such as **Postman**, on your machine.

How to do it...

We will enable our chatroom endpoint to receive messages from the client to print them to the standard output.

Let's start by defining the logger. We will use the logger from the `uvicorn` package (as we did in other recipes – see, for example, *Creating custom middlewares* in *Chapter 8, Advanced Features and Best Practices*), which is the one used by FastAPI as well. In `main.py`, let's write the following:

```
import logging

logger = logging.getLogger("uvicorn")
```

Now let's modify the ws_endpoint function endpoint:

```
@app.websocket("/ws")
async def ws_endpoint(websocket: WebSocket):
    await websocket.accept()
    await websocket.send_text(
        "Welcome to the chat room!"
    )
    while True:
        data = await websocket.receive_text()
        logger.info(f"Message received: {data}")
        await websocket.send_text("Message received!")
```

You might have noticed that we have removed websocket.close() call from the previous recipe and used an infinite while loop. This allows the server side to continuously receive the message from the client and print it to the console without closing the connection. In this case, the connection can be closed only by the client.

This is all you need to read messages from the client and send it to the terminal output.

The server initiates a connection request when the client call the endpoint. With the websocket.receive_text() function, the server opens the connection and it is ready to receive the message from the client. The message is stored into the data variable and it is printed to terminal output. Then the server sends a confirmation message to the client.

Let's test it. Spin up the server by running uvicorn app.main:app from the command line and open Postman. Then apply the following steps.

1. Create a new WebSocket request, and connect to the ws://localhost:8000/ws address.

    Once the connection is established, you will see on the terminal output the message:

    INFO: connection open

    In the response messages of the request you will see:

    Welcome to the chat room! 14:45:19

2. From Postman, you can start sending messages through the **Message** panel below the **URL** field input. For the example try to the send the following message: Hello FastAPI application.

    On the output terminal you will the following message:

    INFO: Message received: Hello FastAPI application

    While in the messages section of the client request you will see the new message:

    Message received! 14:46:20

3.   You can then close the connection from the client by clicking on the **Disconnect** button to the right of the WebSocket **URL** field.

By enabling the server to receive messages from the client, you have just enabled bidirectional communication between the client and server through a WebSocket.

See also

The `Fastapi.WebSocket` instance is, in reality, a `starlette.WebSocket` class from the **Starlette** library. You can do more, such as validating messages as JSON or binary, by using the dedicated methods (the `send_json` or `receive_json` methods).

Check more on the official Starlette documentation page:

- *Starlette Websockets*: `https://www.starlette.io/websockets/`

Handling WebSocket connections and disconnections

When a client establishes a WebSocket connection with a **FastAPI** server, it's crucial to handle the lifecycle of these connections appropriately. This includes accepting incoming connections, maintaining active connections, and handling disconnections gracefully to ensure smooth communication between the client and server. In this recipe, we'll explore how to effectively manage WebSocket connections and gracefully handle disconnections.

Getting ready

To follow the recipe, you will need to have **Postman** or any other tool to test WebSocket connections. Also, you need to already have a WebSocket endpoint implemented in your application. Check the previous two recipes if that is not the case.

How to do it…

We will see how to manage the following two situations:

- Client-side disconnection
- Server-side disconnection

Let's have a look at each of these situations in detail.

Client-side disconnection

You might have noticed in the *Sending and receiving messages over WebSockets* recipe that if the connection is closed from the client (e.g., from Postman) on the server console, a `WebSocketDisconnect` exception propagates, uncaptured.

This is because the disconnection from the client side should be properly handled in a `try-except` block.

Let's adjust the endpoint to take this into account. In the `main.py` module, we modify the `/ws` endpoint as follows:

```
from fastapi.websockets import WebSocketDisconnect


@app.websocket("/ws")
async def ws_endpoint(websocket: WebSocket):
    await websocket.accept()
    await websocket.send_text(
        "Welcome to the chat room!"
    )
    try:
        while True:
            data = await websocket.receive_text()
            logger.info(f"Message received: {data}")
    except WebSocketDisconnect:
        logger.warning(
            "Connection closed by the client"
        )
```

If you rerun the server, connect to the WebSocket endpoint, /ws, and then disconnect, you won't see the error propagation anymore.

Server-side disconnection

In this situation, the connection is closed by the server. Suppose the server will close the connection based on a specific message such as the `"disconnect"` text string, for example.

Let's implement it in the /ws endpoint:

```
@app.websocket("/ws")
async def ws_endpoint(websocket: WebSocket):
    await websocket.accept()
    await websocket.send_text(
        "Welcome to the chat room!"
    )
```

```
while True:
    data = await websocket.receive_text()
    logger.info(f"Message received: {data}")
    if data == "disconnect":
        logger.warn("Disconnecting...")
        await websocket.close()
        break
```

All we need is the checking condition of the `data` string content to then call the `websocket.close` method and exit the `while` loop.

If you run the server, try to connect to the WebSocket /ws endpoint, and send the `"disconnect"` string as a message, the connection will be closed by the server.

You have seen how to manage disconnection and connection handshakes for a WebSocket endpoint, however, we still need to manage the right status code and messages for each. Let's check this in the following recipe.

Handling WebSocket errors and exceptions

WebSocket connections are susceptible to various errors and exceptions that can occur during the lifecycle of a connection. Common issues include connection failures, message parsing errors, and unexpected disconnections. Properly handling errors and correctly communicating with the client is essential to maintaining a responsive and resilient WebSocket-based application. In this recipe, we'll explore how to handle WebSocket errors and exceptions effectively in FastAPI applications.

Getting ready

The recipe will show how to manage WebSocket errors that can happen for a specific endpoint. We will showcase how to improve the /ws endpoint defined in the *Handling WebSocket connections and disconnections* recipe.

How to do it...

The way the /ws endpoint is coded in the previous recipe returns the same response code and message when the server closes the connection. Just like for HTTP responses, FastAPI allows you to personalize the response to return a more meaningful message to the client.

Let's see how to do it. You can use a solution like the following:

```
from fastapi import status


@app.websocket("/ws")
```

```
async def chatroom(websocket: WebSocket):
    if not websocket.headers.get("Authorization"):
        return await websocket.close()

    await websocket.accept()
    await websocket.send_text(
        "Welcome to the chat room!"
    )
    try:
        while True:
            data = await websocket.receive_text()
            logger.info(f"Message received: {data}")
            if data == "disconnect":
                logger.warn("Disconnecting...")
                return await websocket.close(
                    code=status.WS_1000_NORMAL_CLOSURE,
                    reason="Disconnecting...",
                )
    except WebSocketDisconnect:
        logger.warn("Connection closed by the client")
```

We have specified to the `websocket.close` method a status code and reason that will be transmitted to the client.

If we now spin up the server and send the disconnect message from the client, you will see a disconnection log message in the response window, like this:

```
Disconnected from localhost:8000/ws 14:09:08
1000 Normal Closure:  Disconnecting...
```

This is all you need to gracefully disconnect your WebSocket connection.

Alternative solution

Similarly to how an `HTTPException` instance is rendered for HTTP requests (see the *Handling errors and exceptions* recipe in *Chapter 1, First Steps with FastAPI*), FastAPI also enables the use of `WebSocketException` for WebSocket connections, which is rendered automatically as a response.

To better understand, imagine we want to disconnect the client if they write something that isn't allowed – for example, the `"bad message"` text string. Let's modify the chatroom endpoint:

```
@app.websocket("/ws")
async def ws_endpoint(websocket: WebSocket):
    await websocket.accept()
```

```
await websocket.send_text(
    "Welcome to the chat room!"
)
try:
    while True:
        data = await websocket.receive_text()
        logger.info(f"Message received: {data}")
        if data == "disconnect":
            logger.warn("Disconnecting...")
            return await websocket.close(
                code=status.WS_1000_NORMAL_CLOSURE,
                reason="Disconnecting...",
            )
        if "bad message" in data:
            raise WebSocketException(
                code=status.WS_1008_POLICY_VIOLATION,
                reason="Inappropriate message"
            )
except WebSocketDisconnect:
    logger.warn("Connection closed by the client")
```

If you spin up the server and try to send whatever message contains the `"bad message"` string, the client will be disconnected. Furthermore, on the **Response** panel section of Postman of your WebSocket connection you will see the following log message:

```
Disconnected from localhost:8000/ws 14:51:40
1008 Policy Violation: Inappropriate message
```

You have just seen how to communicate WebSocket errors to the client by raising the appropriate exception. You can use this strategy for a variety of errors that can arise while running the application and have to be correctly communicated to the API consumer.

See also

WebSocket is a relatively new protocol compared to HTTP, so it is still evolving with time. Although status codes are not extensively used, like for HTTP, you can find definitions of WebSockets codes at the following links:

- *WebSocket Close Code Number Registry*: `https://www.iana.org/assignments/websocket/websocket.xml#close-code-number`

You can also find a list of the compatibility of WebSocket events for browsers on the following page:

- *WebSocket CloseEvent*: `https://developer.mozilla.org/en-US/docs/Web/API/CloseEvent`

Furthermore, the `WebSocketException` class in FastAPI is documented at the official documentation link:

- *FastAPI WebSocketExcpetion API documentation*: `https://fastapi.tiangolo.com/reference/exceptions/#fastapi.WebSocketException`

Implementing chat functionality with WebSockets

Real-time chat functionality is a common feature in many modern web applications, enabling users to communicate instantly with each other. In this recipe, we'll explore how to implement chat functionality using WebSockets in FastAPI applications.

By leveraging WebSockets, we will create a bidirectional communication channel between the server and multiple clients, allowing messages to be sent and received in real time.

Getting ready

To follow the recipe, you need to have a good understanding of WebSockets and know how to build a WebSocket endpoint using FastAPI.

Additionally, having some basic knowledge of HTML and JavaScript can help create simple web pages for the application. The recipe we'll be using is the foundation of our chat application.

Also, we will use the `jinja2` package to apply basic templating for the HTML page. Make sure to have it in your environment. If you didn't install packages with `requirements.txt`, install `jinja2` with `pip`:

```
$ pip install jinja2
```

Once the installation is complete, we are ready to start with the recipe.

How to do it...

To build the application, we will need to build three core pieces – the WebSocket connections manager, the WebSocket endpoint, and the chat HTML page:

1. Let's start by building the connection manager. The role of the connection manager is to keep track of open WebSocket connections and broadcast messages to active ones. Let's define the ConnectionManager class in a dedicated ws_manager.py module under the app folder:

    ```python
    import asyncio
    from fastapi import WebSocket

    class ConnectionManager:
        def __init__(self):
            self.active_connections: list[WebSocket] = []

        async def connect(self, websocket: WebSocket):
            await websocket.accept()
            self.active_connections.append(websocket)

        def disconnect(self, websocket: WebSocket):
            self.active_connections.remove(websocket)

        async def send_personal_message(
            self, message: dict, websocket: WebSocket
        ):
            await websocket.send_json(message)

        async def broadcast(
            self, message: json, exclude: WebSocket = None
        ):
            tasks = [
                connection.send_json(message)
                for connection in self.active_connections
                if connection != exclude
            ]
            await asyncio.gather(*tasks)
    ```

    The async def connect method will be responsible for the handshake and adding the WebSocket to the list of active ones. The def disconnect method will remove the WebSocket from the list of active connections. The async def send_personal_message method will send a message to a specific WebSocket. Finally, async def broadcast will send the message to all the active connections except one, if specified.

The connection manager will then be used in the chat WebSocket endpoint.

2. Let's create the WebSocket endpoint in a separate module called `chat.py`. Let's initialize the connection manager:

```
from app.ws_manager import ConnectionManager
conn_manager = ConnectionManager()
```

Then we define the router:

```
from fastapi import APIRouter
router = APIRouter()
```

And finally, we can define the WebSocket endpoint:

```
from fastapi import WebSocket, WebSocketDisconnect


@router.websocket("/chatroom/{username}")
async def chatroom_endpoint(
    websocket: WebSocket, username: str
):
    await conn_manager.connect(websocket)
    await conn_manager.broadcast(
        f"{username} joined the chat",
        exclude=websocket,
    )

    try:
        while True:
            data = await websocket.receive_text()
            await conn_manager.broadcast(
                {"sender": username, "message": data},
                exclude=websocket,
            )
            await conn_manager.send_personal_message(
                {"sender": "You", "message": data},
                websocket,
            )
    except WebSocketDisconnect:
        conn_manager.disconnect(websocket)
        await connection_manager.broadcast(
            {
                "sender": "system",
                "message": f"Client #{username} "
                "left the chat",
```

```
      }
    )
```

3. After a new client joins a chat, the connection manager sends a message to all chat participants to notify them of the new arrival. The endpoint uses the `username` path parameter to retrieve the client's name. Don't forget to add the router to the FastAPI object in the `main.py` file:

```
from app.chat import router as chat_router
# rest of the code

app = FastAPI()
app.include_router(chat_router)
```

Once the WebSocket endpoint is ready, we can create the endpoint to return the HTML chat page.

4. The page endpoint will return an HTML page rendered with **Jinja2**.

    The HTML chat page named `chatroom.html` should be stored in a `templates` folder in the project root. We will keep the page simple with the JavaScript tag embedded.

    The HTML part will look like this:

```
<!doctype html>
<html>
  <head>
    <title>Chat</title>
  </head>

  <body>
    <h1>WebSocket Chat</h1>
    <h2>Your ID: <span id="ws-id"></span></h2>
    <form action="" onsubmit="sendMessage(event)">
      <input
        type="text"
        id="messageText"
        autocomplete="off"
      />
      <button>Send</button>
    </form>
    <ul id="messages"></ul>
    <script>
        <!--content of js script -->
    <script/>
  </body>
</html>
```

The `<script>` tag will contain the Javascript code that will connect to the WebSocket /chatroom/{username} endpoint with the client name as a parameter, send the message from the client page, receive messages from the server, and render the message text on the page in the messages list section.

You can find an example in the GitHub repository, in the `templates/chatroom.html` file. Feel free to make your own version or download it.

5. To conclude, we need to build the endpoint that returns the HTML page. We can build it in the same `chat.py` module:

```python
from fastapi.responses import HTMLResponse
from fastapi.templating import Jinja2Templates
from app.ws_manager import ConnectionManager


conn_manager = ConnectionManager()


templates = Jinja2Templates(directory="templates")


@router.get("/chatroom/{username}")
async def chatroom_page_endpoint(
    request: Request, username: str
) -> HTMLResponse:
    return templates.TemplateResponse(
        request=request,
        name="chatroom.html",
        context={"username": username},
    )
```

The endpoint will take as a path parameter the username of the client that will show in the chat conversation.

You have set up a basic chat room within your FastAPI application with the WebSockets protocol. You only have to spin up the server with `uvicorn app.main:app` and connect to `http://localhost:8000/chatroom/your-username` from your browser. Then, from another page, connect to the same address with a different username and start exchanging messages between the two browsers.

How it works...

When connecting to the `GET /chatroom/{username}` endpoint address (`http://localhost:8000/chatroom/{username}`), the server will use the username to render the HTML page customized to the username.

The HTML will contain the code to make the connection to the `/chatroom` WebSocket endpoint and create a new WebSocket connection for each user.

The endpoint will then use the `ConnectionManager()` connection manager object to exchange messages between all clients through the HTML page.

See also

We have used a basic feature of the Jinja2 templating library. However, you can free your creativity and discover the potential of this package by looking at the documentation:

- *Jinja2 Documentation*: `https://jinja.palletsprojects.com/en/3.1.x/`

Optimizing WebSocket performance

WebSocket connections provide a powerful mechanism for real-time communication between clients and servers. To ensure the optimal performance and scalability of WebSocket applications, it's essential to implement effective optimization techniques and a way to measure them. In this recipe, we will see how to benchmark WebSocket endpoints to test the number of connections supported by the connection and suggest practical tips and techniques to optimize WebSocket performance in your FastAPI applications.

Getting ready

Besides knowledge of how to set up a WebSocket endpoint, we will use the *Implementing chat functionality with WebSockets* recipe to benchmark the traffic supported. You can also follow the recipe by applying the strategy to your application.

Whether you apply it to your application or the chat functionality, it can be useful to include some message logs to be printed during the endpoint execution.

For example, for the WebSocket `/chatroom/{username}` endpoint, you can add a log after each message broadcast as follows:

```python
import logging
logger = logging.getLogger("uvicorn")

@router.websocket("/chatroom/{username}")
async def chatroom_endpoint(
    websocket: WebSocket, username: str
):
    await conn_manager.connect(websocket)
    await conn_manager.broadcast(
        # method's parameters
    )

    logger.info(f"{username} joined the chat")
```

```
    try:
        while True:
            data = await websocket.receive_text()
            await conn_manager.broadcast(
                # method's parameters
            )
            await conn_manager.send_personal_message(
                # method's parameters
            )
            logger.info(
                f"{username} says: {data}"
            )
    except WebSocketDisconnect:
        conn_manager.disconnect(websocket)
        await conn_manager.broadcast(
            # method's paramters
        )
        logger.info(f"{username} left the chat")
```

We are now ready to create a benchmark script to test our chat functionality.

How to do it...

Let's create the script file under the root folder and call it benchmark_websocket.py. A typical benchmark script should do the following tasks:

- Define a function to run the FastAPI server
- Define another function to connect *n* number of clients of the WebSocket endpoint and exchange a certain number of messages
- Wrap up the previous steps by running the server in a separate process and running the clients

Here are the steps to create the script:

1. Let's start by defining a function to run our server:

   ```
   import uvicorn
   from app.main import app

   def run_server():
       uvicorn.run(app)
   ```

The `run_server` function is an alternative to the command-line `uvicorn app.main:app` command we are used to running from the terminal.

2. Now let's define a function that will create a certain number of clients that will connect to the WebSocket endpoint and exchange some messages:

```python
import asyncio
from websockets import connect

async def connect_client(
    n: int, n_messages: int = 3
):
    async with connect(
        f"ws://localhost:8000/chatroom/user{n}",
    ) as client:
        for _ in range(n_messages):
            await client.send(
                f"Hello World from user{n}"
            )
            await asyncio.sleep(n * 0.1)
        await asyncio.sleep(2)
```

To simulate concurrent connection patterns, we use an `async def` function. This will enable us to evaluate the server's performance under the high load of simultaneous requests to the endpoint.

Furthermore, we added some asynchronous sleeping time (`asyncio.sleep`) between messages to simulate the human behavior of the chat's client.

3. Then, we can execute all the previous functions in a single overall `async def main` function as follows:

```python
import multiprocessing

async def main(n_clients: int = 10):
    p = multiprocessing.Process(target=run_server)
    p.start()
    await asyncio.sleep(1)

    connections = [
        connect_client(n) for n in range(n_clients)
    ]

    await asyncio.gather(*connections)
```

```
        await asyncio.sleep(1)
        p.terminate()
```

The function creates a process to spin up the server, start it, wait some time to finish the startup, and simultaneously run all the clients to call the server.

4. Finally, to make it run, we need to pass it to the event loop if it is run as a script. We can do it like this:

```
    if __name__ == "__main__":
        asyncio.run(main())
```

To run the script, simply run it as a Python script from the command line:

```
$ python benchmark_websocket.py
```

If you leave the default value for the parameter n_clients, you will probably see all the messages flowing on the terminal. However, by increasing n_clients, depending on your machine, at some point, the script will not be able to run anymore and you will see socket connection errors popping up. That means that you passed the limit to support new connections with your endpoint.

What we did is the core of a basic script to benchmark. You can further expand the script based on your needs by adding timing or parametrization to have a broader view of your application's capabilities.

You can also do the same by using dedicated test frameworks, similar to what we did in the *Performance testing for high traffic applications* recipe in *Chapter 5, Testing and Debugging FastAPI Applications*, for HTTP traffic.

There's more...

Benchmarking your WebSocket is only the first step to optimize your application performance. Here is a checklist of actions that you can take to improve your application performance and reduce errors:

*   **Make unit tests for WebSockets with TestClient**: FastAPI's TestClient also supports WebSocket connections, so use it to ensure that the behavior of the endpoint is the one expected and does not change during the development process.

*   **Handle errors gracefully**: Implement error handling mechanisms to gracefully manage exceptions and errors encountered during WebSocket communication. Use try/except blocks to handle specific error conditions. Also, when possible, use async for over while True when managing message exchanges. This will automatically capture and treat disconnection errors.

*   **Use connection pool managers**: Connection pool managers improve performance and code maintainability when handling multiple clients, such as in chat applications.

See also

You can see more on unit testing WebSockets with FastAPI in the official documentation:

- *Testing WebSockets in FastAPI*: `https://fastapi.tiangolo.com/advanced/testing-websockets/`

Securing WebSocket connections with OAuth2

Securing WebSocket connections is paramount to safeguarding the privacy and security of user interactions in real-time applications. By implementing authentication and access control mechanisms, developers can mitigate risks associated with unauthorized access, eavesdropping, and data tampering. In this recipe, we will see how to create a secure WebSocket connection endpoint with OAuth2 token authorization in your FastAPI applications.

Getting ready

To follow the recipe, you should already know how to set up a basic WebSocket endpoint – explained in the *Setting up WebSockets in FastAPI* recipe in this chapter.

Furthermore, we are going to use **OAuth2** with a password and a bearer token. We will apply the same strategy we used to secure HTTP endpoints in the *Securing your API with OAuth2* recipe in *Chapter 3, Building RESTful APIs with FastAPI*. Feel free to have a look before starting the recipe.

Before starting the recipe, let's create a simple WebSocket endpoint, `/secured-ws`, in the `main.py` module:

```python
@app.websocket("/secured-ws")
async def secured_websocket(
    websocket: WebSocket,
    username: str
):
    await websocket.accept()
    await websocket.send_text(f"Welcome {username}!")
    async for data in websocket.iter_text():
        await websocket.send_text(
            f"You wrote: {data}"
        )
```

The endpoint will accept any connection with a parameter to specify the username. Then it will send a welcome message to the client and return each message received to the client.

The endpoint is insecure since it does not have any protection and can be easily reached. Let's dive into the recipe to see how to protect it with OAuth2 authentication.

How to do it...

At the time of writing, there is no support for the OAuth2PasswordBearer class for WebSocket in FastAPI. This means that checking the bearer token in the headers for WebSocket is not as straightforward as it is for HTTP calls. However, we can create a WebSocket-specific class that is derived from the one used by HTTP to achieve the same functionality as follows.

1. Let's do it in a dedicated module under the app folder called ws_password_bearer.py:

```python
from fastapi import (
    WebSocket,
    WebSocketException,
    status,
)
from fastapi.security import OAuth2PasswordBearer


class OAuth2WebSocketPasswordBearer(
    OAuth2PasswordBearer
):
    async def __call__(
        self, websocket: WebSocket
    ) -> str:
        authorization: str = websocket.headers.get(
            "authorization"
        )
        if not authorization:
            raise WebSocketException(
                code=status.HTTP_401_UNAUTHORIZED,
                reason="Not authenticated",
            )
        scheme, param = authorization.split()
        if scheme.lower() != "bearer":
            raise WebSocketException(
                code=status.HTTP_403_FORBIDDEN,
                reason=(
                    "Invalid authentication "
                    "credentials"
                ),
            )
        return param
```

We will use it to create a get_username_from_token function to retrieve the username from the token. You can create the function in a dedicated module – security.py.

2. Let's define the `oauth2_scheme_for_ws` object:

```
from app.ws_password_bearer import (
    OAuth2WebSocketPasswordBearer,
)


oauth2_scheme_for_ws = OAuth2WebSocketPasswordBearer(
    tokenUrl="/token"
)
```

3. The `tokenUrl` argument specifies the callback endpoint to call to retrieve the token. This endpoint should be built according to the token resolution you use. After that, we can create a function that retrieves the username from the token:

```
def get_username_from_token(
    token: str = Depends(oauth2_scheme_for_ws),
) -> str:
    user = fake_token_resolver(token)
    if not user:
        raise WebSocketException(
            code=status.HTTP_401_UNAUTHORIZED,
            reason=(
                "Invalid authentication credentials"
            )
        )
    return user.username
```

The purpose of the `fake_token_resolver` function is to simulate the process of resolving a token. This function can be found in the `security.py` file in the GitHub repository of the chapter. Furthermore, the example contains only two users, `johndoe` and `janedoe`, who can be used later for testing. Also, the `security.py` module from the GitHub repository contains the `POST /token` endpoint to be used to retrieve the token.

However, it is important to mention that this function does not provide any actual security and it is only used for example purposes. In a production environment, it is recommended to use a **JWT Authorization** token or an external provider for token resolution (see the *Working with OAuth2 and JWT for authentication* and *Using third-party authentication* recipes – both in *Chapter 4, Authentication and Authorization*).

4. Now let's use it to secure our WebSocket endpoint, /secured-ws, in the main.py module:

```
from import Annotated
from fastapi import Depends
from app.security import get_username_from_token

@app.websocket("/secured-ws")
async def secured_websocket(
    websocket: WebSocket,
    username: Annotated[
        get_username_from_token, Depends()
    ]
):
    # rest of the endpoint
```

This is all you need to build a secured WebSocket endpoint.

To test it, spin up the server from the terminal by running the following:

```
$ uvicorn app.main:app
```

When attempting to connect to the WebSocket endpoint using Postman or another tool to the address ws://localhost:8000/secured-ws, an authorization error will occur, and the connection will be rejected before the handshake.

To allow the connection, we need to retrieve the token and pass it through the headers of the WebSocket request in **Postman**. You can retrieve the token from the dedicated endpoint or, if you use the fake token generator from the GitHub repository, you simply append the tokenized string to the username. For example, for johndoe, the token would be tokenizedjohndoe.

Let's pass it through the header. In Postman, you can pass the bearer token to the WebSocket request in the **Headers** tab by adding a new header. The header will have a key called Authorization and value that will be bearer tokenizedjohndoe.

Now, if you try to connect, it should connect and you will be able to exchange messages with the endpoint.

You have just secured a WebSocket endpoint in FastAPI. By implementing OAuth2 authorization, you can enhance the security posture of your FastAPI applications and safeguard WebSocket communication against potential threats and vulnerabilities.

Exercise

Try to build a secure chat functionality where users need to log in to participate in the chat.

Tips: The endpoint that returns the HTML page should check for the bearer token in the cookies. If the cookie is not found or the bearer token is not valid, it should redirect the client to a login page that puts the token in the browser's cookies.

You can use the `response.RedirectResponse` class from the `fastapi` package to handle redirections. The usage is quite straightforward and you can have a look at the documentation page at the link:

`https://fastapi.tiangolo.com/advanced/custom-response/#redirectresponse`.

See also

Integrating **OAuth2** into WebSockets in FastAPI with an `OAuth2PasswordBearer`-like class is a current topic of interest, and it is expected to evolve quickly over time. You can follow the ongoing discussion in the FastAPI GitHub repository:

- *OAuth2PasswordBearer with WebSocket Discussion*: `https://github.com/tiangolo/fastapi/discussions/8983`

10

Integrating FastAPI with other Python Libraries

In this chapter, we will delve into the process of expanding the capabilities of **FastAPI** by integrating it with other **Python** libraries. By harnessing the power of external tools and libraries, you can enhance the functionality of your FastAPI applications and unlock new possibilities for creating dynamic and feature-rich web services.

Throughout this chapter, you will learn how to integrate FastAPI with a diverse range of Python libraries, each serving a different purpose and offering unique functionalities. From taking advantage of advanced natural language processing capabilities with **Cohere** and **LangChain** to integrating real-time communication features with **gRPC** and **GraphQL**, you will discover how to harness the full potential of FastAPI in conjunction with other popular Python tools.

By integrating FastAPI with other Python libraries, you will be able to build sophisticated web applications that go beyond simple **REST APIs**. Whether you are developing a chatbot powered by natural language processing or integrating **machine learning** (**ML**) models for intelligent decision-making, the possibilities are endless.

By the end of this chapter, you will be equipped with the knowledge and skills to effectively leverage external tools and resources, enabling you to build sophisticated and feature-rich APIs that meet the needs of your users.

This chapter includes the following recipes:

- Integrating FastAPI with gRPC
- Connecting FastAPI with GraphQL
- Using ML models with Joblib
- Integrating FastAPI with Cohere
- Integrating FastAPI with LangChain

Technical requirements

To follow the recipes in this chapter, it is important to have a good understanding of FastAPI. Additionally, since this chapter demonstrates how to integrate FastAPI with external libraries, having a basic knowledge of each library can be beneficial.

However, we will provide external links for you to review any of the concepts that are used in the recipes. You can also refer back to this chapter whenever you need to integrate a technology with FastAPI.

The code used in the chapter is hosted on GitHub at `https://github.com/PacktPublishing/FastAPI-Cookbook/tree/main/Chapter10`.

It is recommended to set up a virtual environment for the project in the project root folder to efficiently manage dependencies and maintain project isolation.

For each recipe, you can install all the dependencies at once within your virtual environment by using the `requirements.txt` file provided in the GitHub repository in the following project folder:

```
$ pip install -r requirements.txt
```

Let's start delving into this recipe and discovering the potential of coupling FastAPI with external libraries.

Integrating FastAPI with gRPC

gRPC is a high-performance, open source universal **Remote Procedure Call** (**RPC**) framework originally developed by Google. It is designed to be efficient, lightweight, and interoperable across different programming languages and platforms. Integrating FastAPI with gRPC allows you to leverage the power of RPC for building efficient, scalable, and maintainable APIs.

The recipe will show how to build a gateway between a REST client and a gRPC server by using FastAPI.

Getting ready

To follow the recipe, it can be beneficial to have some previous knowledge of protocol buffers. You can have a look at the official documentation at `https://protobuf.dev/overview/`.

Also, we will use the proto3 version to define the `.proto` files. You can check the language guide at `https://protobuf.dev/programming-guides/proto3/`.

We will create a dedicated root project folder for the recipe called `grpc_gateway`.

Aside from `fastapi` and `uvicorn`, you also need to install the `grpcio` and `grpcio-tools` packages. You can do this by using the `requirements.txt` file provided in the GitHub repository in your environment or by explicitly specifying the packages with the `pip` command in your environment as follows:

```
$ pip install fastapi uvicorn grpcio grpcio-tools
```

Before starting with the recipe, let's build a basic gRPC server with one method that takes a message from the client and sends back a message as well by following these steps.

1. Under the root project let's create a `grpcserver.proto` file containing the definition of our server as follows:

```
syntax = "proto3";

service GrpcServer{

    rpc GetServerResponse(Message)
    returns (MessageResponse) {}

}
```

2. In the same file, we will define the `Message` and `MessageResponse` messages as follows:

```
message Message{
string message = 1;
}

message MessageResponse{
string message = 1;
bool received = 2;
}
```

From the `.proto` file we have just created, we can automatically generate the Python code necessary to integrate the service and gRPC client as well with a proto compiler.

3. Then, from the command line terminal, run the following command:

```
$ python -m grpc_tools.protoc \
--proto_path=. ./grpcserver.proto \
--python_out=. \
--grpc_python_out=.
```

This command will generate two files: grpcserver_pb2_grpc.py and grpcserver_pb2.py. The grpcserver_pb2_grpc.py file contains the class to build the server a support function and a `stub` class that will be used by the client, while the grpcserver_pb2.py module contains the classes that define the messages. In our case, these are `Message` and `MessageResponse`.

4. Now let's write a script to run the gRPC server. Let's create a file called `grpc_server.py` and define the server class as follows:

```python
from grpcserver_pb2 import MessageResponse
from grpcserver_pb2_grpc import GrpcServerServicer


class Service(GrpcServerServicer):
    async def GetServerResponse(
        self, request, context
    ):
        message = request.message
        logging.info(f"Received message: {message}")
        result = (
            "Hello I am up and running, received: "
            f"{message}"
        )
        result = {
            "message": result,
            "received": True,
        }
        return MessageResponse(**result)
```

5. Then we will define the function to run the server on the localhost on port `50015` as follows:

```python
import grpc

from grpcserver_pb2_grpc import (
    add_GrpcServerServicer_to_server
)


async def serve():
    server = grpc.aio.server()
    add_GrpcServerServicer_to_server(
        Service(), server
    )
    server.add_insecure_port("[::]:50051")
    logging.info("Starting server on port 50051")
    await server.start()
    await server.wait_for_termination()
```

6. We then close the script by running the `serve` function into the event loop:

```
import asyncio
import logging

if __name__ == "__main__":
    logging.basicConfig(level=logging.INFO)
    asyncio.run(serve())
```

This is all we need to build the gRPC server. Now we can run the script from the command line:

```
$ python ./grpc_server.py
```

If everything has been set up correctly you will see the following log message on the terminal:

```
INFO:root:Starting server on port 50051
```

With the gRPC server running, we can now create our gateway by leveraging FastAPI.

How to do it...

We will create a FastAPI application with one GET `/grpc` endpoint that will take a message as a parameter, forward the request to the gRPC server, and return the message from the gRPC server to the client. Let's go through the following steps to build a basic gateway application.

1. Under the project root folder, let's create a folder called `app` with a `main.py` module containing the server as follows:

```
from fastapi import FastAPI

app = FastAPI()
```

2. Now let's create the response class schema with Pydantic that will reflect the `MessageResponse` proto class as follows:

```
from pydantic import BaseModel

class GRPCResponse(BaseModel):
    message: str
    received: bool
```

3. Then we will initialize the `grpc_channel` object, which is an abstraction layer for gRPC calls containing the gRPC service URL, as follows:

```
grpc_channel = grpc.aio.insecure_channel(
    "localhost:50051"
)
```

4. Finally, we can create our endpoint as follows:

```
@app.get("/grpc")
async def call_grpc(message: str) -> GRPCResponse:
    async with grpc_channel as channel:
        grpc_stub = GrpcServerStub(channel)
        response = await grpc_stub.GetServerResponse(
            Message(message=message)
        )
        return response
```

Once we have created our FastAPI application, let's spin up the server from the command line:

```
$ uvicorn app.main:app
```

Open the interactive documentation at `http://localhost:8000/docs` and you will see the new endpoint that will take a message parameter and return the response from the gRPC server. If you try to call it, you will also see the log message for the call on the gRPC server terminal.

You have successfully set up a REST-gRPC gateway by using FastAPI!

There's more...

We have created a gateway that supports Unary RPC, which is a simple RPC that resembles a normal function call. It involves sending a single request, which is defined in the `.proto` file, to the server and receiving a single response back from the server. However, there are various types of RPC implementations available that allow for the streaming of messages from the client to the server or from the server to the client, as well as ones that allow for bidirectional communication.

Creating a REST gateway using FastAPI is a relatively easy task. For more information on how to implement different types of gRPC in Python, you can refer to the following article: `https://www.velotio.com/engineering-blog/grpc-implementation-using-python`. Once you have mastered these concepts, you can easily integrate them into FastAPI and build a complete gateway for gRPC services.

See also

You can dive deeper into protocol buffer and how you can use it in Python code in the official documentation:

- *Protocol Buffer Python Generated Code*: `https://protobuf.dev/reference/python/python-generated/`

You check more on how to implement gRPC in Python code at the gRPC official documentation page:

- *gRPC Python Tutorial*: `https://grpc.io/docs/languages/python/basics/`

Also, have a look at the examples on GitHub:

- *gRPC Python GitHub Examples*: `https://github.com/grpc/grpc/tree/master/examples/python`

Connecting FastAPI with GraphQL

GraphQL is a query language for APIs and a runtime for executing queries. It provides an efficient, powerful, and flexible alternative to traditional REST APIs by allowing clients to specify exactly what data they need. Integrating FastAPI with GraphQL enables you to build APIs that are highly customizable and capable of handling complex data requirements. In this recipe, we will see how to connect FastAPI with GraphQL to query a user database, allowing you to create GraphQL schemas, define resolvers, and expose a GraphQL endpoint in your FastAPI application.

Getting ready

To follow the recipe, it can be beneficial to ensure you already have some basic knowledge about GraphQL. You can have a look at the official documentation at `https://graphql.org/learn/`.

In the GitHub repository folder of this chapter, there is a folder named `graphql`, which we will consider as the root project folder. To implement GraphQL, we will be utilizing the Strawberry library. Please ensure that you have it installed in your environment along with FastAPI. You can install it by using the `requirements.txt` file located in the project root of the repository or by using the `pip` command by running the following:

```
$ pip install fastapi uvicorn strawberry-graphql[fastapi]
```

Once the installation is complete, we can start the recipe.

How to do it...

Let's create a basic GraphQL endpoint that retrieves users from a specific country in a database. Let's do it through the following steps.

1. Let's create a `database.py` module containing a list of users that we will use as a database source. Define the `User` class as follows:

    ```
    from pydantic import BaseModel


    class User(BaseModel):
        id: int
        username: str
        phone_number: str
        country: str
    ```

2. You can then write a `users_db` object, which will be a list of `User` class objects, or copy the one from the respective `database.py` file on the GitHub repository at `https://raw.githubusercontent.com/PacktPublishing/FastAPI-Cookbook/main/Chapter10/graphql/database.py`.

    It will look like this:

    ```
    users_db: list[User] = [
        User(
            id=1,
            username="user1",
            phone_number="1234567890",
            country="USA",
        ),
    # other users
    ]
    ```

    We will use this list as a database for our simple query.

3. In a separate module called `graphql_utils.py`, we will define the query. But first, let's define the model returned by the query as follows:

    ```
    import strawberry
    @strawberry.type
    class User:
        username: str
        phone_number: str
        country: str
    ```

4. Then we will define the query as follows:

```
@strawberry.type
class Query:
    @strawberry.field
    def users(
        self, country: str | None
    ) -> list[User]:
        return [
            User(
                username=user.username,
                phone_number=user.phone_number,
                country=user.country,
            )
            for user in users_db
            if user.country == country
        ]
```

The query takes a country as an argument and returns all the users for the country.

5. Now, in the same file, let's create the GraphQL schema with the FastAPI router:

```
from strawberry.fastapi import GraphQLRouter
schema = strawberry.Schema(Query)
graphql_app = GraphQLRouter(schema)
```

The last line will create a `fastapi.Router` instance that will handle the endpoint.

6. Let's finalize the endpoint by adding the router to the main FastAPI instance in a separate `main.py` module as follows:

```
from fastapi import FastAPI
from graphql_utils import graphql_app

app = FastAPI()
app.include_router(graphql_app, prefix="/graphql")
```

We added the endpoint to the FastAPI instance and defined the `/graphql` path.

This is all you need to setup a GraphQL endpoint within FastAPI.

To explore the potential of the endpoint, let's run the server from the command line:

```
$ uvicorn main:app
```

Now let's open the browser at `http://localhost:8000/graphql`. You will see an interactive page for the endpoint. The page is divided into two panels. The left contains the query editor and the right will show you the response.

Try to make the following GraphQL query:

```
{
  users(country: "USA") {
    username
    country
    phoneNumber
  }
}
```

You will see the result on the right panel, which will look something like this:

```
{
  "data": {
    "users": [
      {
        "username": "user1",
        "country": "USA",
        "phoneNumber": "1234567890"
      }
    ]
  }
}
```

You have learned how to create an interactive GraphQL endpoint. By combining RESTful endpoints with GraphQL, the potential for data querying and modification can be greatly expanded. Real-world scenarios may involve using REST endpoints to modify the database by adding, modifying, or removing records. GraphQL can then be used to query the database and extract valuable insights.

See also

You can consult the FastAPI official documentation on how to integrate GraphQL:

- *FastAPI GraphQL Documentation*: `https://fastapi.tiangolo.com/how-to/graphql/`

Also, in the Strawberry documentation, you can find a dedicated page on FastAPI integration:

- *Integrate FastAPI with Strawberry*: `https://strawberry.rocks/docs/integrations/fastapi`

Using ML models with Joblib

ML models are powerful tools for data analysis, prediction, and decision-making in various applications. FastAPI provides a robust framework for building web services, making it an ideal choice for deploying ML models in production environments. In this recipe, we will see how to integrate an ML model with FastAPI using **Joblib**, a popular library for model serialization and deserialization in Python.

We will develop an AI-powered doctor application that can diagnose diseases by analyzing the symptoms provided.

> **Warning**
> Note that the diagnoses provided by the AI doctor should not be trusted in real-life situations, as it is not reliable.

Getting ready

Prior knowledge of ML is not mandatory but having some can be useful to help you follow the recipe.

We will apply the recipe to a new project, so create a folder named `ai_doctor` that we will use as the project root folder.

To ensure that you have all the necessary packages in your environment, you can install them using the `requirements.txt` file provided in the GitHub repository or from the command line:

```
$ pip install fastapi[all] joblib scikit-learn
```

We will download the model from the Hugging Face Hub, a centralized hub hosting pre-trained ML models that are ready to be used.

We will use the `human-disease-prediction` model, which is a relatively lightweight linear logistic regression model developed with the `scikit-learn` package. You can check it out at the following link: `https://huggingface.co/AWeirdDev/human-disease-prediction`.

To download it, we will leverage the provided `huggingface_hub` Python package, so make sure you have it in your environment by running the following:

```
$ pip install huggingface_hub
```

Once the installation is complete, we can proceed with building our AI doctor.

How to do it...

Let's follow these steps to create our AI doctor:

1. Let's start by writing the code to accommodate the ML model. In the project root folder, let's create the app folder containing a module called utils.py. In the module, we will declare a symptoms_list list containing all the symptoms accepted by the model. You can download the file directly from the GitHub repository at https://raw.githubusercontent. com/PacktPublishing/FastAPI-Cookbook/main/Chapter10/ai_doctor/ app/utils.py.

    You can find the complete list on the model's documentation page at https://huggingface. co/AWeirdDev/human-disease-prediction.

2. Still in the app folder, let's create the main.py module that will contain the FastAPI server class object and the endpoint. To incorporate the model into our application, we will utilize the FastAPI lifespan feature.

    We can define the lifespan context manager as follows:

    ```
    from fastapi import FastAPI
    from contextlib import asynccontextmanager
    ml_model = {}
    REPO_ID = "AWeirdDev/human-disease-prediction"
    FILENAME = "sklearn_model.joblib"


    @asynccontextmanager
    async def lifespan(app: FastAPI):
        ml_model["doctor"] = joblib.load(
            hf_hub_download(
                repo_id=REPO_ID, filename=FILENAME
            )
        )

        yield
        ml_model.clear()
    ```

    The lifespan context manager serves as middleware and carries out operations before and after server start and shutdown. It retrieves the model from the Hugging Face Hub and stores it in the ml_model dictionary, so it to be used across the endpoints without the need to reload it every time it is called.

3. Once it has been defined, we need to pass it to the `FastAPI` object class as follows:

```
app = FastAPI(
    title="AI Doctor",
    lifespan=lifespan
)
```

4. Now we need to create the endpoint that will take the symptoms as parameters and return the diagnosis.

The idea is to return each symptom as a path parameter. Since we have 132 possible symptoms, we will create the parameters object dynamically with Pydantic and restrict our model to the first ten symptoms. In the `main.py` file, let's create the `Symptoms` class used to accept the parameters with the `pydantic.create_model` function as follows:

```
from pydantic import create_model
from app.utils import symptoms_list

query_parameters = {
    symp: (bool, False)
    for symp in symptoms_list[:10]
}

Symptoms = create_model(
    "Symptoms", **query_params
)
```

We now have all that we need to create our GET `/diagnosis` endpoint.

5. Let's create our endpoint as follows:

```
@app.get("/diagnosis")
async def get_diagnosis(
    symptoms: Annotated[Symptoms, Depends()],
):
    array = [
        int(value)
        for _, value in symptoms.model_dump().items()
    ]
    array.extend(
        # adapt array to the model's input shape
        [0] * (len(symptoms_list) - len(array))
    )
    len(symptoms_list)
```

```
        diseases = ml_model["doctor"].predict([array])
        return {
            "diseases": [disease for disease in diseases]
        }
```

To test it, as usual, spin up the server with `uvicorn` from the command line by running the following:

```
$ uvicorn app.main:app
```

Open the interactive documentation from the browser at `http://localhost:8000/docs`. You will see the only `GET /diagnosis` endpoint and you will be able to select the symptoms. Try to select some of them and get your diagnosis from the AI doctor you have just created.

You have just created a FastAPI application that integrates an ML model. You can use the same model for different endpoints, but you can also integrate multiple models within the same application with the same strategy.

See also

You can check the guidelines on how to integrate an ML model into FastAPI on the official documentation page:

- *Lifespan Events*: `https://fastapi.tiangolo.com/advanced/events/?h=machine+learning#use-case`

You can have a look at the Hugging Face Hub platform documentation at the link:

- *Hugging Face Hub Documentation*: `https://huggingface.co/docs/hub/index`

Take a moment to explore the capabilities of the `scikit-learn` package by referring to the official documentation:

- *Scikit-learn Documentation*: `https://scikit-learn.org/stable/`

Integrating FastAPI with Cohere

Cohere offers powerful language models and APIs that enable developers to build sophisticated AI-powered applications capable of understanding and generating human-like text.

State-of-the-art language models, such as the **Generative Pre-trained Transformer** (**GPT**) series, have revolutionized how machines comprehend and generate natural language. These models, which are trained on vast amounts of text data, deeply understand human language patterns, semantics, and context.

By leveraging Cohere AI's models, developers can empower their applications to engage in natural language conversations, answer queries, generate creative content, and perform a wide range of language-related tasks.

In this recipe, we will create an AI-powered chatbot using FastAPI and Cohere that suggests Italian cuisine recipes based on user queries.

Getting ready

Before starting the recipe, you will need a Cohere account and an API key.

You can create your account at the page `https://dashboard.cohere.com/welcome/login` by clicking the **Sign up** button at the top. At the time of writing, you can create an account by using your existing GitHub or Google account.

Once logged in, you will see a welcome page and a platform menu on the left with some options. Click on **API keys** to access the API menu.

By default, you will have a trial key that is free of charge, but it is rate limited and it cannot be used for commercial purposes. For the recipe, it will be largely sufficient.

Now create the project root folder called `chef_ai` and store your API key in a file called `.env` under the project root folder as follows:

```
COHERE_API_KEY="your-cohere-api-key"
```

> **Warning**
>
> If you develop your project with a versioning system control such as Git, for example, make sure to not track any API keys. If you have done this already, even unintentionally, revoke the key from the Cohere API keys page and generate a new one.

Aside from the API key, make sure that you also have all the required packages in your environment. We will need `fastapi`, `uvicorn`, `cohere`, and `python-dotenv`. This last package will enable importing environment variables from the `.env` file.

You can install all the packages with the `requirements.txt` file provided in the GitHub repository in the `chef_ai` project folder by running the following:

```
$ pip install -r requirements.txt
```

Alternatively you can install them one by one:

```
$ pip install fastapi uvicorn cohere python-dotenv
```

Once the installation is complete, we can dive into the recipe and create our "chef de cuisine" assistant.

How to do it...

We will create our chef cuisine assistant by using a message completion chat. Chat completion models take a list of messages as input and return a model-generated message as output. The first message to provide is the **system message**.

A system message defines how a chatbot behaves in a conversation, such as adopting a specific tone or acting as a specialist such as a senior UX designer or software engineer. In our case, the system message will tell the chatbot to behave like a chef de cuisine.

Let's create an endpoint to call our chat through the following steps:

1. Let's create a `handlers.py` module under the project root and import the Cohere API key from the `.env` file:

    ```
    from dotenv import load_dotenv
    load_dotenv()
    ```

2. Let's write the system message as follows:

    ```
    SYSTEM_MESSAGE = (
        "You are a skilled Italian top chef "
        "expert in Italian cuisine tradition "
        "that suggest the best recipes unveiling "
        "tricks and tips from Grandma's Kitchen"
        "shortly and concisely."
    )
    ```

3. Define the Cohere asynchronous client as follows:

    ```
    from cohere import AsyncClient
    client = AsyncClient()
    ```

4. Before creating the function the generate the message, let's import the required modules as:

    ```
    from cohere import ChatMessage
    from cohere.core.api_error import ApiError
    from fastapi import HTTPException
    ```

5. Then, we can define the function to generate our message:

```
async def generate_chat_completion(
    user_query=" ", messages=[]
) -> str:
    try:
        response = await client.chat(
            message=user_query,
            model="command-r-plus",
            preamble=SYSTEM_MESSAGE,
            chat_history=messages,
        )
        messages.extend(
            [
                ChatMessage(
                    role="USER", message=user_query
                ),
                ChatMessage(
                    role="CHATBOT",
                    message=response.text,
                ),
            ]
        )
        return response.text

    except ApiError as e:
        raise HTTPException(
            status_code=e.status_code, detail=e.body
        )
```

The function will take in input the user query and the messages previously exchanged during the conversation. If the response is returned with no errors, the messages list is updated with the new interaction, otherwise an HTTPException error is raised.

We utilized **Command R+** model, Cohere's most recent large language model at the time this was written, designed for conversational interactions and long-context tasks.

6. In a new main.py module, located under the project root folder, we can start defining the messages list in the application state at the startup with the lifespan context manager:

```
from contextlib import asynccontextmanager
from fastapi import FastAPI


@asynccontextmanager
```

```
async def lifespan(app: FastAPI):
    yield {"messages": []}
```

7. We then pass the `lifespan` context manager to the app object as:

```
app = FastAPI(
    title="Chef Cuisine Chatbot App",
    lifespan=lifespan,
)
```

8. Finally, we can create our endpoint as follows:

```
from typing import Annotated
from fastapi import Body, Request
from handlers import generate_chat_completion

@app.post("/query")
async def query_chat_bot(
    request: Request,
    query: Annotated[str, Body(min_length=1)],
) -> str:
    answer = await generate_chat_completion(
        query, request.state.messages
    )
    return answer
```

We enforce a minimum length for the query message (`Body(min_length=1)`) to prevent the model from returning an error response.

You have just created an endpoint that interacts with our chef de cuisine chatbot.

To test it, spin up the server with `uvicorn`:

$ uvicorn main:app

Open the interactive documentation and start testing the endpoint. For example, you can prompt the model with a message such as the following:

```
"Hello, could you suggest a quick recipe for lunch to be prepared in
less than one hour?"
```

Read the answer, then try asking the bot to replace some ingredients and continue the chat. Once you have completed your recipe, enjoy your meal!

> **Exercise**
>
> We have created a chatbot endpoint to interact with our assistant. However, for real-life applications, it can be useful to have an endpoint that returns all the messages exchanged.
>
> Create a GET /messages endpoint that returns all the messages in a formatted way.
>
> Also create an endpoint POST /restart-conversation that will flush all the messages and restart the conversation without any previous messages.

See also

You can have a look at the Cohere quickstart on building a chatbot on the official documentation page:

- *Building a Chatbot*: https://docs.cohere.com/docs/building-a-chatbot

In production environment, depending on the project's needs and budget, you might want to choose from the several models available. You can see an overview of the models provided by Cohere here:

- *Models Overview*: https://docs.cohere.com/docs/models

Integrating FastAPI with LangChain

LangChain is a versatile interface for nearly any **Large Language Model** (**LLM**) that allows developers to create LLM applications and integrate them with external data sources and software workflows. It was launched in October 2022 and quickly became a top open source project on GitHub.

We will use LangChain and FastAPI to create an AI-powered assistant for an electronic goods store that provides recommendations and helps users.

We will set up a **Retrieval-Augmented Generation** (**RAG**) application, which involves empowering the model with personalized data to be trained. In this particular case, that would be a document of frequently asked questions.

This recipe will guide you through the process of integrating FastAPI with LangChain to create dynamic and interactive AI assistants that enhance the customer shopping experience.

Getting ready

Before starting the recipe, you will need a Cohere API key. If you don't have it, you can check the *Getting ready* section of the *Integrating FastAPI with Cohere* recipe.

Create a project directory called ecotech_RAG and place the API key within a .env file, labeled as COHERE_API_KEY.

Previous knowledge of LLM and RAG is not required but having it would help.

Aside from the `fastapi` and `uvicorn` packages, you will need to install `python-dotenv` and the packages related to LangChain. You can do this by using `requirements.txt` or by installing them with `pip` as follows:

```
$ pip install fastapi uvicorn python-dotenv
$ pip install langchain
$ pip install langchain-community langchain-cohere
$ pip install chromadb unstructured
```

Once the installation is complete, we can start building our AI shop assistant.

How to do it...

We are going to create an application with a single endpoint that interacts with an LLM from Cohere.

The idea behind LangChain is to provide a series of interconnected modules, forming a chain to establish a workflow linking the user query with the model output.

We will split the process of creating the endpoint to interact with the RAG AI assistant into the following steps:

1. Defining the prompts

2. Ingesting and vectorizing the documents

3. Building the model chain

4. Creating the endpoint

Let's start building our AI-powered assistant.

Defining the prompts

Like for the previous recipe, we will utilize a chat model that takes a list message as input. For this specific use case, however, we will supply the model with two messages: the system message and the user message. LangChain includes template objects for specific messages. Here are the steps to set up our prompts:

1. Under the root project, create a module called `prompting.py`. Let's start the module by defining a template message that will be used as the system message:

```
template: str = """
    You are a customer support Chatbot.
    You assist users with general inquiries
    and technical issues.
    You will answer to the question:
    {question}
    Your answer will only be based on the knowledge
```

```
of the context below you are trained on.
-----------
{context}
-----------
if you don't know the answer,
you will ask the user
to rephrase the question or
redirect the user the support@ecotech.com
always be friendly and helpful
at the end of the conversation,
ask the user if they are satisfied
with the answer if yes,
say goodbye and end the conversation
"""
```

This is a common prompt for customer assistants that contains two variables: `question` and `context`. Those variables will be required to query the model.

2. With that template, we can define the system message as follows:

```
from langchain.prompts import (
    SystemMessagePromptTemplate,
)
system_message_prompt = (
    SystemMessagePromptTemplate.from_template(
        template
    )
)
```

3. The user message does not require specific context and can be defined as follows:

```
from langchain.prompts import (
    HumanMessagePromptTemplate,
)

human_message_prompt = (
    HumanMessagePromptTemplate.from_template(
        template="{question}",
    )
)
```

4.   Then we can group both messages under the dedicated chat message `template` object as follows:

```
from langchain.prompts import ChatPromptTemplate

chat_prompt_template = (
    ChatPromptTemplate.from_messages(
        [system_message_prompt, human_message_prompt]
    )
)
```

This is all we need to set up the prompt object to query our model.

Ingesting and vectorizing the documents

Our assistant will answer user questions by analyzing the documents we will provide to the model. Let's create a `docs` folder under the project root that will contain the documents. First, download the `faq_ecotech.txt` file from the GitHub repository in the `ecotech_RAG/docs` project folder and save it in the local `docs` folder.

You can download it directly at `https://raw.githubusercontent.com/PacktPublishing/FastAPI-Cookbook/main/Chapter10/ecotech_RAG/docs/faq_ecotech.txt`.

Alternatively, you can create your own FAQ file. Just ensure that each question and answer is separated by one empty line.

The information contained in the file will be used by our assistant to help the customers. However, to retrieve the information, we will need to split our documents into chunks and store them as vectors to optimize searching based on similarity.

To split the documents, we will use a character-based text splitter. To store chunks, we will use Chroma DB, an in-memory vector database.

Then, let's create a `documents.py` module that will contain the `load_documents` helper function that will upload the files into a variable as follows:

```
from langchain.text_splitter import (
    CharacterTextSplitter,
)
from langchain_core.documents.base import Document
from langchain_community.document_loaders import (
    DirectoryLoader,
)
from langchain_community.vectorstores import Chroma

async def load_documents(
```

```
    db: Chroma,
):
    text_splitter = CharacterTextSplitter(
        chunk_size=100, chunk_overlap=0
    )

    raw_documents = DirectoryLoader(
        "docs", "*.txt"
    ).load()

    chunks = text_splitter.split_documents(
        raw_documents
    )
    await db.aadd_documents(chunks)
```

The `DirectoryLoader` class uploads the content of all the `.txt` files from the `docs` folder, then the `text_splitter` object reorganizes the documents into document chunks of `100` characters that will be then added to the `Chroma` database.

By utilizing the vectorized database alongside the user query, we can retrieve the relevant context to feed into our model, which will analyze the most significant portion.

We can write a function for this called `get_context` as follows:

```
def get_context(
    user_query: str, db: Chroma
) -> str:
    docs = db.similarity_search(user_query)
    return "\n\n".join(
        doc.page_content for doc in docs
    )
```

The documents have to be stored and vectorized in numerical representations called embedding. This can be done with Chroma, an AI-native vector database.

Then, through a similarity search operation (`db.similaratiry_search`) between the user query and the document chunks, we can retrieve the relevant content to pass as context to the model.

We have now retrieved the context to provide in the chat model system message template.

Building the model chain

Once we have defined the mechanism to retrieve the context, we can build the chain model. Let's build it through the following steps:

1. Let's create a new module called `model.py`. Since we will use Cohere, we will upload the environment variables from the `.env` file with the `dotenv` package as follows:

    ```
    from dotenv import load_dotenv
    load_dotenv()
    ```

2. Then we will define the model we are going to use:

    ```
    from langchain_cohere import ChatCohere

    model = ChatCohere(model="command-r-plus")
    ```

    We will use the same module we used in the previous recipe, Command R+.

3. Now we can gather the pieces we have created to leverage the power of LangChain by creating the chain pipeline to query the model as follows:

    ```
    from langchain.schema import StrOutputParser
    from prompting import chat_prompt_template

    chain = (
        chat_prompt_template | model | StrOutputParser()
    )
    ```

We will use the chain object to create our endpoint to expose through the API.

Creating the endpoint

We will make the `app` object instance with the endpoint in the `main.py` module under the project root folder. As always, let's follow these steps to create it:

1. The operation of loading the documents can be quite CPU-intensive, especially in real-life applications. Therefore, we will define a lifespan context manager to execute this process only at server startup. The `lifespan` function will be structured as follows:

    ```
    from contextlib import asynccontextmanager
    from fastapi import FastAPI
    from langchain_cohere import CohereEmbeddings
    from langchain_community.vectorstores import Chroma
    ```

```python
from documents import load_documents

@asynccontextmanager
async def lifespan(app: FastAPI):
    db = Chroma(
        embedding_function=CohereEmbeddings()
    )
    await load_documents(db)
    yield {"db": db}
```

2. We can then pass it to the FastAPI object as follows:

```python
app = FastAPI(
    title="Ecotech AI Assistant",
    lifespan=lifespan
)
```

3. Now, we can define a POST /message endpoint as follows:

```python
from typing import Annotated
from fastapi import Body, Request
from documents import get_context
from model import chain

@app.post("/message")
async def query_assistant(
    request: Request,
    question: Annotated[str, Body()],
) -> str:
    context = get_context(question, request.state.db)
    response = await chain.ainvoke(
        {
            "question": question,
            "context": context,
        }
    )
    return response
```

4. The endpoint will accept a body string text as input and will return the response from the model as a string based on the documents provided in the docs folder at startup.

To test it, you can spin up the server from the following command:

```
$ uvicorn main:app
```

Once the server has started, open the interactive documentation at http://localhost:8000/docs and you will see the POST /message endpoint we just created.

Try first to send a message that is not related to the assistance, something like the following:

```
"What is the capital of Belgium ?"
```

You will receive an answer like this:

```
"I apologize, but I cannot answer that question as it is outside of
my knowledge base. I am an FAQ chatbot trained to answer specific
questions related to EcoTech Electronics, including our product
compatibility with smart home systems, international shipping costs,
and promotions for first-time customers. If you have any questions
related to these topics, I'd be happy to help! Otherwise, for general
inquiries, you can reach out to our support team at support@ecotech.
com. Is there anything else I can assist you with today regarding
EcoTech Electronics?"
```

Then try to ask, for example, the following:

```
"What kind of payments do you accept?"
```

You will get your assistance answer, which should be something like this:

```
"We want to make sure your shopping experience with us is as smooth
and secure as possible. For online purchases, we currently accept
major credit cards: Visa, Mastercard, and American Express. You also
have the option to pay through PayPal, which offers an additional
layer of security and convenience. \n\nThese payment methods are
integrated into our straightforward online checkout process, ensuring
a quick and efficient transaction. \n\nAre there any specific payment
methods you are interested in using, or do you have any further
questions about our accepted forms of payment? We want to ensure your
peace of mind and a great overall experience shopping with us. \n\nAre
you satisfied with the answer?"
```

You can double check that the answer is in line with what is written in the FAQ document in the docs folder.

You have just implemented a RAG AI-powered assistant with LangChain and FastAPI. You will now be able to implement your own AI assistant for your application.

> **Exercise**
>
> We have implemented the endpoint to interact with the chat model that will answer based on the document provided. However, real-life API applications will allow the addition of new documents interactively.
>
> Create a new `POST /document` endpoint that will add a file in the `docs` folder and reload the documents in the code.
>
> Have a look at the *Working with file uploads and downloads* recipe in *Chapter 2, Working with Data*, to see how to upload files in FastAPI.

See also

You can have a look at the quickstart in the LangChain documentation:

* *LangChain Quickstart*: `https://python.langchain.com/v0.1/docs/get_started/quickstart/`

We have used Chroma, a vector database largely used for ML applications. Feel free to have a look at the documentation:

* *Chroma*: `https://docs.trychroma.com/`

11
Middleware and Webhooks

In this chapter, we delve into the advanced and crucial aspects of middleware and webhooks in FastAPI. Middleware in FastAPI allows you to process requests and responses globally before they reach your route handlers and after they leave them. Webhooks, on the other hand, enable your FastAPI application to communicate with other services by sending real-time data updates. Both middleware and webhooks are essential for building robust, efficient, and scalable applications.

We will start by exploring how to create custom **Asynchronous Server Gateway Interface (ASGI)** middleware from scratch. This will give you a deep understanding of how middleware works at a fundamental level.

Next, we'll develop middleware specifically for response modification, allowing you to intercept and alter responses before they are sent back to the client.

We will also cover handling **Cross-Origin Resource Sharing (CORS)** with middleware. This is particularly important for applications that need to interact with different domains securely. Finally, we will dive into creating webhooks in FastAPI, demonstrating how to set them up and test them effectively.

By the end of this chapter, you will have a comprehensive understanding of how to implement and utilize middleware and webhooks in your FastAPI applications. These skills will enable you to build more dynamic, responsive, and integrated web services.

In this chapter, we're going to go through the following recipes:

- Creating custom ASGI middleware
- Developing middleware for request modification
- Developing middleware for response modification
- Handling CORS with middleware
- Restricting incoming requests from hosts
- Implementing webhooks

Technical requirements

At this stage of the book, you should already have a good understanding of the basics of FastAPI, how to install it, and how to run it.

The code used in the chapter is hosted on GitHub at the following address: `https://github.com/PacktPublishing/FastAPI-Cookbook/tree/main/Chapter11`.

It is recommended to set up a virtual environment for the project in the project root folder to efficiently manage dependencies and maintain project isolation.

Throughout the chapter, we will only be using the standard `fastapi` library with `uvicorn`. You can install all the dependencies within your virtual environment using `pip` from the command line by running the following:

```
$ pip install fastapi uvicorn
```

For the *Handling CORS with middleware* recipe, having some basic knowledge of JavaScript and HTML will be beneficial.

Creating custom ASGI middleware

ASGI is a specification for Python web servers and applications to communicate with each other, designed to support asynchronous functionality. Middleware is a critical component in web applications, providing a way to process requests and responses.

We have already seen, in the *Creating custom middleware* recipe in *Chapter 8, Advanced Features and Best Practices*, how to create custom middleware. However, this technique relies on the `BasicHTTPMiddleware` class from the Starlette library, which is a high-level implementation of HTTP middleware.

In this recipe, we'll learn how to create custom ASGI middleware from scratch and integrate it into a FastAPI application. The middleware will be simple and will only print log message information on the terminal.

This approach provides greater control over the request/response cycle compared to the `BasicHTTPMiddleware` class, allowing for advanced customizations and the creation of any kind of middleware with a deeper level of customization.

Getting ready

Since we will use the Starlette library to build middleware, a sound knowledge of this library would be beneficial, although not necessary.

Regarding the development environment, we will exclusively utilize the `fastapi` package with `uvicorn`. Ensure they are installed in your environment.

How to do it...

Let's start by creating a project root folder called `middleware_project`. Under the root folder, create a folder called `middleware` containing a module called `asgi_middleware.py`. Let's start the module by declaring the logger that we will use during the middleware call:

```python
import logging
logger = logging.getLogger("uvicorn")
```

Then we can define the middleware class as follows:

```python
from starlette.types import (
    ASGIApp, Scope, Receive, Send
)

class ASGIMiddleware:
    def __init__(
        self, app: ASGIApp, parameter: str = "default"
    ):
        self.app = app
        self.parameter = parameter

    async def __call__(
        self,
        scope: Scope,
        receive: Receive,
        send: Send,
    ):
        logger.info("Entering ASGI middleware")
        logger.info(
            f"The parameter is: {self.parameter}"
        )
        await self.app(scope, receive, send)
        logger.info("Exiting ASGI middleware")
```

Then, we need to include the middleware in our application. Under the project root folder, create the `main.py` module containing the FastAPI class to run the application, as follows:

```python
from fastapi import FastAPI
from starlette.middleware import Middleware
from middleware.asgi_middleware import ASGIMiddleware

app = FastAPI(
```

```
        title="Middleware Application",
        middleware=[
            Middleware(
                ASGIMiddleware,
                parameter="example_parameter",
            ),
        ]
    )
```

This is all you need to implement custom ASGI middleware in a FastAPI application.

How it works...

To see the middleware in action, let's create a general endpoint in main.py module, as in the following example:

```
@app.get("/")
async def read_root():
    return {"Hello": "Middleware World"}
```

Spin up the server by running uvicorn main:app from the command line. You will see the following messages:

```
INFO:      Started server process [2064]
INFO:      Waiting for application startup.
INFO:      Entering ASGI middleware
INFO:      The parameter is: example_parameter
```

Among the messages, you will notice those indicating that we have already entered the middleware. Now try to call the root endpoint. You can do it by opening the browser at http://localhost:8000/.

Still on the terminal, this time you will notice both middleware messages for entering and exiting:

```
INFO:      Entering ASGI middleware
INFO:      The parameter is: example_parameter
INFO:      127.0.0.1:55750 - "GET / HTTP/1.1" 200 OK
INFO:      Exiting ASGI middleware
```

As we understand from the logs, we entered the middleware twice, once on the startup and once when calling the endpoint, but we exited the middleware only once.

This is why the ASGI middleware intercepts every event of the application, not only the HTTP request but also the lifespan event, which includes the startup and shutdown.

Information on the event type within the middleware is stored in the `scope` parameter of the `__call__` method. Let's include the following logs in the `ASGIMiddleware.__call__` method to improve our understanding of the mechanism:

```
class ASGIMiddleware:
    def __init__(
        self,
        app: ASGIApp,
        parameter: str = "default",
    ):
    # method content
    async def __call__(
        self,
        scope: Scope,
        receive: Receive,
        send: Send,
    ):
        # previous logs
        logger.info(
            f"event scope: {scope.get('type')}"
        )
        await self.app(scope, receive, send)
```

If you restart the server and remake the call to `http://localhost:8000/`, you will now see the log messages specifying the event scope type to be `lifespan` at the server startup and `http` after the endpoint call.

There's more...

We have just seen how to build ASGI middleware as a class. However, you can also do it by leveraging the function decorator pattern. For example, you can build the same middleware like this:

```
def asgi_middleware(
    app: ASGIApp, parameter: str = "default"
):
    @functools.wraps(app)
    async def wrapped_app(
        scope: Scope, receive: Receive, send: Send
    ):
        logger.info(
            "Entering second ASGI middleware"
        )
        logger.info(
```

```
                    f"The parameter you proved is: {parameter}"
            )
            logger.info(
                f"event scope: {scope.get('type')}"
            )
            await app(scope, receive, send)
            logger.info("Exiting second ASGI middleware")
    return wrapped_app
```

This is the equivalent of the `ASGIMiddleware` class defined earlier in the *How to do it…* subsection. To make it work, it should be passed as an argument to the FastAPI instance in exactly the same way:

```
from middleware.asgi_middleware import asgi_middleware

app = FastAPI(
    title="Middleware Application",
    middleware=[
        Middleware(
            asgi_middleware,
            parameter="example_parameter",
        ),
    ]
)
```

Based on your personal preference, you can choose the style you prefer. However, for the rest of the chapter, we will continue using the middleware class style.

See also

You can read more on the ASGI specification in the dedicated documentation page:

- *ASGI Documentation*: `https://asgi.readthedocs.io/en/latest/`

Middleware classes in FastAPI derive from the Starlette library. You can find extensive documentation on creating ASGI middleware on the Starlette documentation page:

- *Pure ASGI Middleware*: `https://www.starlette.io/middleware/#pure-asgi-middleware`

Developing middleware for request modification

Middleware in web applications serves as a powerful tool for processing requests. Custom middleware can intercept and modify these messages, allowing developers to add or modify functionalities.

In this recipe, we'll focus on developing custom ASGI middleware to modify responses before they are sent to the client by hashing the body of each request, if necessary. This approach provides the flexibility to add or change response headers, body content, and other properties dynamically. By the end of the recipe, you will be able to develop custom middleware to control every API request.

Getting ready

Before we begin, please make sure you have completed the previous recipe, *Creating custom ASGI middleware*, to create specific custom ASGI middleware. We will be working on the middleware_project application, but the recipe can easily be applied to any application.

Before creating the middleware, in the main.py module, let's create a POST /send endpoint that accepts body content in the request, as follows:

```
import logging
logger = logging.getLogger("uvicorn")

@app.post("/send")
async def send(message: str = Body()):
    logger.info(f"Message: {message}")
    return message
```

The endpoint will print the body content to the terminal and return it as a response as well.

Now that we have our endpoint, we can create the middleware to hash the body content before sending it to the endpoint.

How to do it...

In the middleware folder, let's create a module called request_middleware.py that will host our middleware class. Let's go through the following steps to create the middleware:

1. Start the module with the required imports like this:

    ```
    from starlette.types import (
        ASGIApp, Scope, Receive, Send, Message,
    )
    from hashlib import sha1
    ```

    We will use the types from the Starlette library to create the middleware class and the sha1 function to hash the body.

2. Given that only certain HTTP verbs accept the body (POST and PUT, but not GET for example), we will pass to the middleware the paths as parameters where the modifications should be applied.

   Create a middleware class called HashBodyContentMiddleware, as follows:

```python
class HashBodyContentMiddleWare:
    def __init__(
        self, app: ASGIApp, allowed_paths: list[str]
    ):
        self.app = app
        self.allowed_paths = allowed_paths
```

   We will pass the list of paths into the allowed_paths parameters.

3. Define the __call__ method of the class:

```python
async def __call__(
    self,
    scope: Scope,
    receive: Receive,
    send: Send,
):
    if (
        scope["type"] != "http"
        or scope["path"]
        not in self.allowed_paths
    ):
        await self.app(scope, receive, send)
        return
```

   If the event is not an HTTP request or the path is not listed, the middleware won't take any action and will leave the request passing through the next step.

4. The information about the body is brought by the receive variable. However, the receive variable is a coroutine, and it should be passed as that to the self.app object. We will overcome this by creating a new coroutine within the function, as follows:

```python
# continue the __call__ method content
async def receive_with_new_body() -> Message:
    message = await receive()
    assert message["type"] == "http.request"

    body = message.get("body", b"")
    message["body"] = (
        f'"{sha1(body).hexdigest()}"'.encode()
```

```
        )

        return message

    await self.app(
        scope,
        receive_with_new_body,
        send,
    )
```

The body request will be modified by the coroutine that is passed to the following steps of the FastAPI object application.

5. Now we need to add the middleware to the FastAPI instance. We can do it in the `main.py` module. But this time we will leverage the `add_middleware` method of the FastAPI instance object like this:

```
app.add_middleware(
    HashBodyContentMiddleWare,
    allowed_paths=["/send"],
)
```

Now the application will make the request pass through our middleware.

This is all you need to implement it. To test the middleware, let's spin up the server with `uvicorn` from the command line by running the following:

```
$ uvicorn main:app
```

Then go to the interactive documentation at `http://localhost:8000/docs` and test the `POST/send` endpoint. For example, check whether you can send a body string like this:

```
"hello middleware"
```

If everything is correctly done, you should receive a response body like this:

```
"14bb256ec4a292037c01bdbdd3eac61f328515f3"
```

You have just implemented custom ASGI middleware that hashes the body for the specified endpoints.

This was a simple example, but the potential of controlling requests is limitless. For example, you can use it to introduce an additional security layer to prevent cross-scripting injection of undesired content.

See also

Creating middleware to modify the request is documented on the Starlette documentation page:

- *Inspecting or modifying the request*: `https://www.starlette.io/middleware/#inspecting-or-modifying-the-request`

Developing middleware for response modification

Besides processing requests, middleware in web applications is also a powerful tool for processing responses. Custom middleware allows us to intercept responses before they are returned to the API caller. This can be useful for checking response content or personalizing the response. In this recipe, we will develop custom ASGI middleware to add customized headers to all the responses.

Getting ready

We will be creating custom ASGI middleware that modifies the response of each HTTP call. Before we get started on this recipe, take a look at the *Creating custom ASGI middleware* recipe. Also, this recipe will be complementary to the previous recipe, *Developing middleware for request modification*.

While you can apply this recipe to your own project, we will continue working on the `middleware_project` project that we initialized in the *Developing middleware for request modification* recipe.

How to do it...

We will create our middleware class in a dedicated module in the `middleware` folder. We will call the module `response_middleware.py`. Let's start building the middleware by going through the following steps.

1. Let's start writing the imports we will use to define the middleware:

```
from typing import Sequence
from starlette.datastructures import MutableHeaders
from starlette.types import (
    ASGIApp, Receive, Scope, Send, Message
)
```

2. Then, we can start defining the `ExtraHeadersResponseMiddleware` middleware class as follows:

```
class ExtraHeadersResponseMiddleware:
    def __init__(
        self,
        app: ASGIApp,
        headers: Sequence[tuple[str, str]],
    ):
        self.app = app
        self.headers = headers
```

3. We will pass the headers list as an argument to the middleware. Then, the `__call__` method will be as follows:

```
async def __call__(
    self,
    scope: Scope,
    receive: Receive,
    send: Send,
):
    if scope["type"] != "http":
        return await self.app(
            scope, receive, send
        )
```

4. We restrain the middleware to HTTP event calls. Similar to what we saw in the previous recipe, *Developing middleware for request modification*, we modify the send object, which is a coroutine, and we pass it to the next middleware, as follows:

```
async def send_with_extra_headers(
    message: Message
):
    if (
        message["type"]
        == "http.response.start"
    ):
        headers = MutableHeaders(
            scope=message
        )
        for key, value in self.headers:
            headers.append(key, value)

    await send(message)
```

```
        await self.app(
            scope, receive, send_with_extra_headers
        )
```

The response's headers are generated from the `message` parameter of the `send_with_extra_headerds` coroutine object.

5. Once the middleware is defined, we need to add it to the `FastAPI` object instance to make it effective. We can add it in the `main.py` module as follows:

```
app.add_middleware(
    ExtraHeadersResponseMiddleware,
    headers=(
        ("new-header", "fastapi-cookbook"),
        (
            "another-header",
            "fastapi-cookbook",
        ),
    ),
)
```

Here, we add two headers to the response, `new-header` and `another-header`.

To test it, spin up the server by running `uvicorn main:app` and open the interactive documentation. Call one of the endpoints and check the headers in the response.

Here is the list of the headers you get when calling the `GET /` endpoint:

```
another-header: fastapi-cookbook
content-length: 28
content-type: application/json
date: Thu,23 May 2024 09:24:41 GMT
new-header: fastapi-cookbook
server: uvicorn
```

You will find the two headers we previously added to the default ones.

You have just implemented middleware that modifies API responses.

See also

In the Starlette documentation, you can find an example of how to create middleware that modifies the response:

- *Inspecting or modifying the response*: `https://www.starlette.io/middleware/#inspecting-or-modifying-the-response`

Handling CORS with middleware

CORS is a security feature implemented in web browsers to prevent malicious websites from making unauthorized requests to APIs hosted on different origins. When building APIs, especially for public consumption, it's crucial to handle CORS properly to ensure legitimate requests are served while unauthorized ones are blocked.

In this recipe, we will explore how to handle CORS using custom middleware in FastAPI. This approach allows us to deeply understand the CORS mechanism and gain flexibility in customizing the behavior to fit specific requirements.

Getting ready

We will apply the recipe to the `middleware_project` application. Make sure you have the FastAPI application running with at least the `GET /` endpoint already set up.

Since the recipe will show how to set up CORS middleware to manage CORS, you will need a simple HTML web page that calls our API.

You can create one yourself or download the `cors_page.html` file from the project's GitHub repository. The file is a simple HTML page that sends a request to the FastAPI application at `http://localhost:8000/` and displays the response on the same page.

Before starting the recipe, spin up your FastAPI application by running `uvicorn main:app`. To view the page, open `cors_page.html` using a modern browser. Then, open the developer console. In most browsers, you can do this by right-clicking on the page, selecting **Inspect** from the menu, and then toggling to the **Console** tab.

On the page, press the **Send CORS Request** button. You should see an error message on the command line like the following:

```
Access to fetch at 'http://localhost:8000/' from origin 'null' has
been blocked by CORS policy: Response to preflight request doesn't
pass access control check: No 'Access-Control-Allow-Origin' header is
present on the requested resource. If an opaque response serves your
needs, set the request's mode to 'no-cors' to fetch the resource with
CORS disabled.
```

That means that the call has been blocked by the CORS policy.

Let's start the recipe and see how to fix it.

How to do it...

In FastAPI, CORS can be handled with a dedicated `CORSMiddleware` class from the Starlette library.

Let's add the middleware to our application in the `main.py` module:

```
from fastapi.middleware.cors import CORSMiddleware
# rest of the module
app.add_middleware(
    CORSMiddleware,
    allow_origins=["*"],
    allow_methods=["*"],
    allow_headers=["*"],
)
```

Now, rerun the server, open `cors_page.html` again, and try to press the **Send CORS Request** button. This time, you see the response message directly on the page.

The `allow_origins` parameter specifies the host origin from which the CORS should be allowed. If `allow_origins=[*]`, it means that any origin is allowed.

The `allow_methods` parameter specifies the HTTP methods that are allowed. By default, only `GET` is allowed, and if `allow_methods=[*]`, it means that all methods are allowed.

Then, the `allow_headers` parameter specifies the headers that are allowed. Similarly, if we use `allow_headers=[*]`, it means that all headers are allowed.

In a production environment, it's important to carefully evaluate each of these parameters to ensure security standards and to make your application run safely.

This is all that's needed to implement CORS middleware for allowing CORS from clients.

See also

For more information about CORS, check out the **Mozilla** documentation page:

- *CORS*: `https://developer.mozilla.org/en-US/docs/Web/HTTP/CORS`

You can see more about the functionalities and discover other parameters of the CORS middleware in FastAPI on the documentation page:

- *Use CORSMiddleware*: `https://fastapi.tiangolo.com/tutorial/cors/#use-corsmiddleware`

You can also have a look at the Starlette documentation page:

- *CORSMiddleware*: `https://www.starlette.io/middleware/#corsmiddleware`

Restricting incoming requests from hosts

In modern web applications, security is paramount. One crucial aspect of security is ensuring that your application only processes requests from trusted sources. This practice helps to mitigate risks such as **Domain Name System** (**DNS**) rebinding attacks, where an attacker tricks a user's browser into interacting with an unauthorized domain.

FastAPI provides middleware called `TrustedHostMiddleware`, which allows you to specify which hosts are considered trusted. Requests from any other hosts will be rejected. This recipe will guide you through setting up and using the `TrustedHostMiddleware` class to secure your FastAPI application by accepting requests only from specific hosts.

Getting ready

We will apply the recipe to the `middleware_project` application. The application will need to be working with at least one endpoint to test.

How to do it...

Let's restrict the request to calls coming from localhost. In `main.py`, let's import `TrustedHostMiddleware` and add it to the FastAPI object instance application, as follows:

```
from fastapi.middleware.trustedhost import (
    TrustedHostMiddleware,
)

# rest of the module

app.add_middleware(
    TrustedHostMiddleware,
    allowed_hosts=["localhost"],
)
```

To test it, let's try to refuse a call. Let's spin up the server by broadcasting our service to the network. We can do it by specifying the undefined host address, `0.0.0.0`, when running `uvicorn`, as follows:

```
$ uvicorn main:app --host=0.0.0.0
```

This will make our application visible to the network.

To retrieve the address of your machine within the local network, you can run `ipconfig` on Windows or `ip addr` on Linux or macOS.

From another device connected to the same local network as the machine running our FastAPI application (such as a smartphone), open a browser and enter `http://<your local address>:8000`. If everything is correctly set up, you will see the following message in the browser:

```
Invalid host header
```

While on the machine running the FastAPI server, you will see a log message like the following:

```
INFO: <client ip>:57312 - "GET / HTTP/1.1" 400 Bad Request
```

This is all you need to set up middleware to prevent your application from being reached by undesired hosts.

See also

You can learn more about `TrustedHostMiddleware` on the FastAPI documentation page:

- *TrustedHostMiddleware*: `https://fastapi.tiangolo.com/advanced/middleware/#trustedhostmiddleware`

Since `TrustedHostMiddleware` is defined in the Starlette library, you can also find it in the Starlette documentation at the following link:

- *TrustedHostMiddleware*: `https://www.starlette.io/middleware/#trustedhostmiddleware`

Implementing webhooks

Webhooks play a crucial role in modern web development by enabling different systems to communicate and respond to events in real time. They are essentially HTTP callbacks triggered by specific events in one system, which then send a message or payload to another system. This asynchronous event-driven architecture allows for seamless integration with third-party services, real-time notifications, and automated workflows. Understanding how to implement webhooks effectively will empower you to build more interactive and responsive applications.

In this recipe, we will see how to create webhooks in FastAPI. We will create a webhook that notifies the webhook subscribers for each request of the API, acting like a monitoring system. By the end of this recipe, you will be able to implement a robust webhook system in your FastAPI application, facilitating real-time communication and integration with other services.

Getting ready

To set up the webhook for sending requests to the subscriber, we will use custom ASGI middleware. Please ensure that you have already followed the *Creating custom ASGI middleware* recipe. We will be continuing our work on the `middleware_project` API. However, you will find guidelines on how to implement your webhook that can be easily adapted to the specific needs of your project.

If you are starting a new project from scratch, make sure to install the `fastapi` package with `uvicorn` in your environment. You can do this using `pip`:

```
$ pip install fastapi uvicorn
```

Once you have the packages, we can start the recipe.

How to do it...

To build a webhook system in our API, we will need to do the following:

1. Set up the URL registration system.
2. Implement the webhook callbacks.
3. Document the webhook.

Let's go through the implementation.

Setting up the URL registration system

A webhook call will send an HTTP request to the list of URLs registered to the webhook. The API will require a URL registration system. This can be achieved by creating a dedicated endpoint that will store the URL in a stateful system, such as a database. However, for demonstration purposes, we will store the URLs in the application state, which might also be a good choice for small applications.

Let's create it by going through the following steps:

1. In `main.py`, let's create the lifespan context manager to store the registered URLs:

```
from contextlib import asynccontextmanager

@asynccontextmanager
async def lifespan(app: FastAPI):
    yield {"webhook_urls": []}
```

2.   Let's pass the lifespan as an argument to the FastAPI object, as follows:

```
app = FastAPI(
    lifespan=lifespan,
# rest of the parameters
)
```

3.   Then, we can create the endpoint to register the URL, as follows:

```
@app.post("/register-webhook-url")
async def add_webhook_url(
    request: Request, url: str = Body()
):
    if not url.startswith("http"):
        url = f"http://{url}"
    request.state.webhook_urls.append(url)
    return {"url added": url}
```

The endpoint will accept a text string in the body. If the http or https protocol is missing in the string, an "http://" string will be prepended to the URL before being stored.

You have just implemented the URL registration system. Now, let's continue to implement the webhook callbacks.

Implementing the webhook callbacks

After setting up the registration system, we can begin creating the webhook's calls. As previously stated, this particular webhook will alert subscribers for every API call. We'll utilize this information to develop specialized middleware that will handle the calls. Let's do it by following these steps:

1.   Let's create a new module in the middleware folder called webhook.py and define the event to communicate with the subscribers:

```
from pydantic import BaseModel
class Event(BaseModel):
    host: str
    path: str
    time: str
    body: str | None = None
```

2. Then, we define a coroutine that will be used to make the requests to the subscriber URLs, as follows:

```python
import logging
from httpx import AsyncClient
client = AsyncClient()

logger = logging.getLogger("uvicorn")

async def send_event_to_url(
    url: str, event: Event
):
    logger.info(f"Sending event to {url}")
    try:
        await client.post(
            f"{url}/fastapi-webhook",
            json=event.model_dump(),
        )
    except Exception as e:
        logger.error(
            "Error sending webhook event "
            f"to {url}: {e}"
        )
```

The client sends a request to the URL. If the request fails, a message is printed to the terminal.

3. We then define the middleware that will intercept the request. We start with the imports, as follows:

```python
from asyncio import create_task
from datetime import datetime
from fastapi import Request
from starlette.types import (
    ASGIApp, Receive, Scope, Send,
)
```

We then add the `WebhookSenderMiddleware` class, as follows

```python
class WebhookSenderMiddleWare:
    def __init__(self, app: ASGIApp):
        self.app = app

    async def __call__(
        self,
```

```
        scope: Scope,
        receive: Receive,
        send: Send,
    ):
```

4. We will filter only the HTTP requests, as follows:

```
if scope["type"] == "http":
    message = await receive()
    body = message.get("body", b"")
    request = Request(scope=scope)
```

5. We continue in the same __call__ function by defining the event object to pass to the webhook subscribers:

```
event = Event(
    host=request.client.host,
    path=request.url.path,
    time=datetime.now().isoformat(),
    body=body,
)
```

6. Then, we iterate the calls over the URLs by running the send_event_to_url coroutine, as follows:

```
urls = request.state.webhook_urls
for url in urls:
    await create_task(
        send_event_to_url(url, event)
    )
```

7. We finalize the method by returning the modified receive function to the application:

```
async def continue_receive():
    return message

await self.app(
    scope, continue_receive, send
)
return

await self.app(scope, receive, send)
```

We have just defined the middleware that will make the calls.

8. Now we need to import the `WebhookSenderMiddleWare` middleware in the application. We can do this inside `main.py` as follows:

```
from middleware.webhook import (
WebhookSenderMiddleWare
)
# rest of the code
app.add_middleware(WebhookSenderMiddleWare)
```

The application will now include our middleware to handle the webhook callbacks.

That is all you need to implement a complete webhook within your FastAPI application.

Documenting the webhook

It is important to provide API users with documentation on how the webhook functions. FastAPI allows us to document a webhook in the OpenAPI documentation.

To accomplish this, you need to create a function with an empty body and declare it as a webhook endpoint. You can do it in `main.py` as well:

```
@app.webhooks.post("/fastapi-webhook")
def fastapi_webhook(event: Event):
    """_summary_

    Args:
        event (Event): Received event from webhook
        It contains information about the
        host, path, timestamp and body of the request
    """
```

You can also provide an example of the body content by adding specifications to the `Event` class in the `middleware/webhook.py` module, as follows:

```
class Event(BaseModel):
    host: str
    path: str
    time: str
    body: str | None = None

    model_config = {
        "json_schema_extra": {
            "examples": [
```

```
        {
            "host": "127.0.0.1",
            "path": "/send",
            "time": "2024-05-22T14:24:28.847663",
            "body": '"body content"',
        }
    ]
  }
}
```

After starting the server with the `uvicorn main:app` command and opening the browser at `http://localhost:8000/docs`, you will find the documentation for `POST /fastapi-webhook` in the **Webhook** section. This documentation explains the call that the API will make to the provided URLs through the `POST register-webhook-url` endpoint.

How it works...

To test the webhook, you can set up a simple server running locally on a specific port. You can create one yourself or download the `http_server.py` file from the GitHub repository. This server will run on port `8080`.

Once you have set up the server, you can run it from the command line:

```
$ python ./http_server.py
```

Leave the server running and make sure the FastAPI application is running on a separate terminal.

Open the interactive documentation at `http://localhost:8000/docs`. Using the `POST /register-webhook-url` endpoint, add the `"localhost:8080"` address. Make sure you specify the correct port in the URL.

Now try to call any of the endpoints to the API. The FastAPI application will make a call to the server listening at port `8080`. If you check the service terminal, you will see the messages streaming on the terminal containing the information for each call.

There's more...

While the basic implementation of webhooks is powerful, several advanced concepts and enhancements can make your webhook system more robust, secure, and efficient. Some of the most relevant ones are as follows:

- **Authentication**: To ensure that your API can securely communicate with a webhook endpoint, you can implement any sort of authentication, from API to OAuth.

- **Retry mechanism**: Webhooks rely on HTTP, which is not always reliable. There may be instances where the webhook delivery fails due to network issues, server downtime, or other transient errors. Implementing a retry mechanism ensures that webhook events are eventually delivered even if the initial attempt fails.

- **Persistent storage**: Storing webhook events in a database allows you to keep an audit trail, troubleshoot issues, and replay events if necessary. You can use SQLAlchemy, a SQL toolkit and **object-relational mapping** library for Python, to save webhook events in a relational database.

- **WebSocket webhook**: For real-time updates, you can set up a WebSocket server in FastAPI and notify clients through WebSocket connections when webhooks are received.

- **Rate limiting**: To prevent abuse and server overload, rate limiting can be applied to the webhook endpoint. This ensures that a single client cannot overwhelm the server with too many requests in a short period.

Webhooks are crucial for constructing interactive, event-driven applications that seamlessly integrate with third-party systems. Utilize them to their fullest potential.

See also

If you want to learn more about webhook applications, check out the **Red Hat** blog page explaining what it is and how it is used in modern applications:

- *What is a webhook?*: `https://www.redhat.com/en/topics/automation/what-is-a-webhook`

You can also refer to the FastAPI documentation for information on how to document webhook endpoints in the OpenAPI documentation:

- *OpenAPI Webhooks*: `https://fastapi.tiangolo.com/advanced/openapi-webhooks/`

12

Deploying and Managing FastAPI Applications

In this chapter, we delve into the essential aspects of deploying and managing FastAPI applications. As you develop your FastAPI projects, understanding how to effectively run, secure, and scale them is crucial for ensuring performance and reliability in production environments. This chapter will equip you with the knowledge and tools needed to deploy your FastAPI applications seamlessly, leveraging various technologies and best practices.

You will learn how to utilize the **FastAPI CLI** to run your server efficiently, enabling **HTTPS** to secure your applications, and containerizing your FastAPI projects with **Docker**. Additionally, we will explore techniques for scaling your applications across multiple workers, packaging your applications for distribution, and deploying them on cloud platforms such as **Railway**. Each recipe in this chapter provides step-by-step instructions, practical examples, and insights into optimizing your deployment workflow.

By the end of this chapter, you will be proficient in deploying FastAPI applications using modern tools and methodologies. You'll be able to always secure your applications with HTTPS, run them in Docker containers, scale them with multiple workers, and deploy them on the cloud. These skills are invaluable for any developer aiming to take their FastAPI applications from development to production.

In this chapter, we're going to cover the following recipes:

- Running the server with the FastAPI CLI
- Enabling HTTPS on FastAPI applications
- Running FastAPI applications in Docker containers
- Running the server across multiple workers
- Deploying your FastAPI application on the cloud
- Shipping FastAPI applications with Hatch

Technical requirements

This chapter is for advanced users who want to learn how to deploy their FastAPI applications on the cloud. If you are new to FastAPI or Python, you might want to check out the first two chapters of the book.

You can find the chapter's code on GitHub here: https://github.com/PacktPublishing/FastAPI-Cookbook/tree/main/Chapter12.

To manage dependencies and isolate the project, set up a virtual environment in the project root folder.

For the *Running FastAPI applications in Docker containers* and *Running the server across multiple workers* recipes, we will be using Docker. Make sure to install it on your machine.

Running the server with the FastAPI CLI

The FastAPI **command-line interface (CLI)** is a program that runs in the command line. You can use the $ fastapi command to run a FastAPI application, manage a FastAPI project, and do other things. This feature was added in version 0.111.0 recently.

In this recipe, we'll explore how to run a FastAPI application using the FastAPI CLI. This approach can streamline your development workflow and provide a more intuitive way to manage your server.

Getting ready

To run the recipe, ensure you have a minimum FastAPI module with the application with at least one endpoint. We will work on a new application called Live Application, so create a new project folder called live_application with an app subfolder containing a main.py module as follows:

```
from fastapi import FastAPI


app = FastAPI(title="FastAPI Live Application")


@app.get("/")
def read_root():
    return {"Hello": "World"}
```

Also, make sure you have a version of FastAPI higher than 0.111.0 in your environment by running the following from the command line:

```
$ pip install "fastapi~=0.111.0"
```

If you already have installed it, make sure to have the latest version of `fastapi` in your environment. You can do it by running the following:

```
$ pip install fastapi --upgrade
```

Once the installation or the upgrade is completed, we can start the recipe.

How to do it...

With your application set up, simply run the following from the command line:

```
$ fastapi dev
```

You will see detailed information printed on the terminal. Let's check the most important ones.

The first message is like this:

```
INFO     Using path app\main.py
```

In the `fastapi dev` command, we didn't specify an `app.main:app` argument as we used to do with the `uvicorn` command. The FastAPI CLI automatically detects the `FastAPI` object class in the code according to a set of default paths.

The following messages are about the building of the server by looking at the packages and modules to be considered. Then, it explicitly shows the resolved import for the `FastAPI` object class:

```
╭─ Python module file ─╮
│                      │
│    main.py           │
│                      │
╰──────────────────────╯


INFO     Importing module main
INFO     Found importable FastAPI app

╭─ Importable FastAPI app ─╮
│                          │
│    from main import app  │
│                          │
╰──────────────────────────╯


INFO     Using import string main:app
```

Then, you will see messages specifying the running mode with the main addresses similar to this one:

```
———————————  FastAPI CLI - Development mode  ———————————

   Serving at: http://127.0.0.1:8000

   API docs: http://127.0.0.1:8000/docs

   Running in development mode, for production use:

   fastapi run

```

This message indicates that the application is operating in development mode.

This means that it will restart the server automatically when there are code updates, and the server will run on the local address 127.0.0.1.

You can alternatively run the server in production mode by running the following:

```
$ fastapi run
```

This won't apply any reload and the server will make the application visible to the local network of the machine hosting the server.

These are some of the basic commands that you can use to run your FastAPI application with different settings and options. For more advanced features and configurations, you can refer to the FastAPI documentation.

There's more...

The FastAPI CLI relies on the uvicorn command to run. Some of the arguments are similar. For instance, if we want to run the service on a different port number than 8000, we can use the --port parameter, or to specify the host address, we can use --host. You can use the --help parameter to see the command-line documentation with the list of all the available parameters. For example, you can run the following:

```
$ fastapi run --help
```

As an example, to run the application visible to the network, you can pass the unspecified address 0.0.0.0 to the host as follows:

```
$ fastapi run
```

This is the equivalent of the following:

```
$ uvicorn app.main:app --host 0.0.0.0
```

Your application will now be visible to the hosting local network.

See also

You can check more on the functionalities of the FastAPI CLI on the official documentation page:

- *FastAPI CLI*: `https://fastapi.tiangolo.com/fastapi-cli/`

Enabling HTTPS on FastAPI applications

Web applications need security, and **Hypertext Transfer Protocol Secure (HTTPS)** is a basic way to secure communication between clients and servers.

HTTPS scrambles the data sent over the network, preventing unauthorized access and modification.

In this recipe, we will learn how to enable HTTPS on FastAPI applications for local testing. We'll use `mkcert` to make a **Secure Sockets Layer/Transport Layer Security (SSL/TLS)** certificate for local development and give some advice for production deployment. By the end of the recipe, you'll be able to protect your FastAPI application with HTTPS, improving its security and reliability.

Getting ready

Some background information about HTTPS and SSL/TLS certificates can help with this recipe. From a consumer perspective, you can find a good overview at this link: `https://howhttps.works/`.

We will also use an existing application as an example. You can either apply the recipe to your own application or use `Live Application` as a reference.

You will also need `mkcert`, so install it correctly on your machine. Installation depends on your operating system, and you can see the instructions here: `https://github.com/FiloSottile/mkcert?tab=readme-ov-file#installation`.

After installing, run this command from your terminal to see how to use it and check that it works:

```
$ mkcert
```

When the installation is complete, we can start the recipe.

How to do it...

Let's set up our certificates through the following steps.

1. Let's start by allowing our browser to trust certificates created locally with mkcert. Run this simple command:

```
$ mkcert -install
```

You will get a message like this:

```
The local CA is now installed in the system trust store! ⚡
```

This command has added a local certificate in your operating system trust store so that your browsers will automatically accept it as a reliable source of certificates.

2. We can then create the certificates and the private key that the server will use for some domain ranges by running the following:

```
$ mkcert localhost 127.0.0.1
```

This command will generate two files: example.com+5-key.pem for the key and example.com+5.pem for the certificate.

> **Warning**
> To ensure security, do not include certificates and keys in your Git history when you create them. Add the *.pem file extension to the .gitignore file

3. We will have to give the key and the certificate to the server when it starts. At the time of writing, the fastapi command does not support the arguments to pass the key and the certificate to the server, so we will start the server with uvicorn by running the following:

```
$ uvicorn app.main:app --port 443 \
--ssl-keyfile example.com+5-key.pem \
--ssl-certfile example.com+5.pem
```

This command will start the server with the certificate and the key.

This is all you need to set up an HTTPS server connection.

To test it, open your browser, and go to the localhost address.

You will see the lock icon on the address bar, which means that the connection is HTTPS.

However, if you try to reach the address with an HTTP connection at http://localhost:443, you will get an error response.

You can fix this by adding automatic redirection to the HTTPS of the server by using a dedicated middleware provided by FastAPI. Change the `main.py` file as follows:

```
from fastapi import FastAPI
from fastapi.middleware.httpsredirect import (
    HTTPSRedirectMiddleware,
)

app = FastAPI(title="FastAPI Live Application")
app.add_middleware(HTTPSRedirectMiddleware)
# rest of the module
```

Then, restart the server. If you try to connect to `localhost` with an HTTP connection, (for example, `http://localhost:443`), it will automatically redirect you to an HTTPS connection, `https://localhost`. However, since it does not support port redirection, you have to specify port 443 anyway.

You have just enabled an HTTPS connection for your FastAPI application within the server. By enabling HTTPS for your FastAPI application, you have taken an important step toward enhancing web security and user experience. You can now enjoy the features of FastAPI with more confidence and trust.

There's more...

We have seen how to generate TLS/SSL certificates for local testing. In a production environment, it will be similar with the difference that this will involve the **Domain Name System (DNS)** hosting provider.

Here are general guidelines on how to do it:

1. Generate a private key and a **certificate signing request (CSR)** for your domain name. Use tools such as **OpenSSL** or **mkcert** as well. Keep the private key secret. The CSR has information about your domain name and organization that a **certificate authority (CA)** will verify.

2. Submit the CSR to a CA and get a signed certificate. A CA is a trusted entity that issues and validates TLS/SSL certificates. There are self-signed, free, or paid CAs. You may need to provide more proof of your identity and domain ownership depending on the CA. Some popular CAs are **Let's Encrypt**, **DigiCert**, and **Comodo**.

3. Install the certificate and the private key on your web server. The procedure may differ based on the server software and the operating system. You may also need to install intermediate certificates from the CA. Configure your web server to use HTTPS and redirect HTTP to HTTPS.

Often, your hosting service provider may handle the TLS/SSL certificates and configuration for you. Some providers use tools such as **Certbot** to get and renew certificates from Let's Encrypt, or they use their own CA. Check with your provider to see whether they offer such options and how to use them.

See also

The GitHub repository at the following link shows you more possibilities of mkcert:

- *mkcert:* `https://github.com/FiloSottile/mkcert`

In the FastAPI official documentation, you can have a look at HTTPS functioning on the page:

- *About HTTPS:* `https://fastapi.tiangolo.com/deployment/https/`

Instructions on how to run uvicorn in HTTPS mode can be found at the following link:

- *Running with HTTPS:* `https://www.uvicorn.org/deployment/#running-with-https`

You can find details on HTTPSRedirectMiddle on the official documentation page at this link:

- *HTTPSRedirectMiddleware:* `https://fastapi.tiangolo.com/advanced/middleware/#httpsredirectmiddleware`

Running FastAPI applications in Docker containers

Docker is a useful tool that lets developers wrap applications with their dependencies into a container. This method makes sure that the application operates reliably in different environments, avoiding the common *works on my machine* issue. In this recipe, we will see how to make a Dockerfile and run a FastAPI application inside a Docker container. By the end of this guide, you will know how to put your FastAPI application into a container, making it more flexible and simpler to deploy.

Getting ready

You will benefit from some knowledge of container technology, especially Docker, to follow the recipe better. But first, check that **Docker Engine** is set up properly on your machine. You can see how to do it at this link: `https://docs.docker.com/engine/install/`.

If you use Windows, it is better to install **Docker Desktop**, which is a Docker virtual machine distribution with a built-in graphical interface.

Whether you have Docker Engine or Docker Desktop, make sure the daemon is running by typing this command:

```
$ docker images
```

If you don't see any error about the daemon, that means that Docker is installed and working on the machine. The way to start the Docker daemon depends on the installation you choose. Look at the related documentation to see how to do it.

You can use the recipe for your applications or follow along with the `Live Application` application that we introduced in the first recipe, which we are using throughout the chapter.

How to do it...

It is not very complicated to run a simple FastAPI application in a Docker container. The process consists of three steps:

1. Create the Dockerfile.
2. Build the image.
3. Generate the container.

Then, you just have to run the container to have the application working.

Creating the Dockerfile

The Dockerfile contains the instructions needed to build the image from an operating system and the file we want to specify.

It is good practice to create a separate Dockerfile for the development environment. We will name it `Dockerfile.dev` and place it under the project root folder.

We start the file by specifying the base image, which will be as follows:

```
FROM python:3.10
```

This will pull an image from the Docker Hub, which already comes with Python 3.10 integrated. Then, we create a folder called `/code` that will host our code:

```
WORKDIR /code
```

Next, we copy `requirements.txt` into the image and install the packages inside the image:

```
COPY ./requirements.txt /code/requirements.txt
RUN pip install --no-cache-dir -r /code/requirements.txt
```

The `pip install` command runs with the `--no-cache-dir` parameter to avoid `pip` caching operations that wouldn't be beneficial inside a container. Also, in a production environment, for larger applications, it is recommended to pin fixed versions of the packages in `requirements.txt` to avoid potential compatibility issues due to package upgrades.

Then, we can copy the `app` folder containing the application into the image with the following command:

```
COPY ./app /code/app
```

Finally, we define the server startup instruction as follows:

```
CMD ["fastapi", "run", "app/main.py", "--port", "80"]
```

This is all we need to create our `Dockerfile.dev` file.

Building the image

Once we have `Dockerfile.dev`, we can build the image. We can do it by running the following from the command line at the project root folder level:

```
$ docker build -f Dockerfile.dev -t live-application .
```

Since we named our Dockerfile `Dockerfile.dev`, we should specify it in an argument. Once the build is finished, you can check that the image has been correctly built by running the following:

```
$ docker images live-application
```

You should see the details of the image on the output print like this:

```
REPOSITORY         TAG      IMAGE ID      CREATED         SIZE
live-application  latest   7ada80a535c2  43 seconds ago  1.06GB
```

With the image built, we can proceed with creating the container creation.

Creating the container

To create the container and run it; simply run the following:

```
$ docker run -p 8000:80 live-application
```

This will create the container and run it. We can see the container by running the following:

```
$ docker ps -a
```

Since we didn't specify a container name, it will automatically affect a fancy name. Mine, for example, is `bold_robinson`.

Open the browser on `http://localhost:8000` and you will see the home page response of our application.

This is all you need to run a FastAPI application inside a Docker container. Running a FastAPI application in a Docker container is a great way to use the advantages of both technologies. You can easily scale, update, and deploy your web app with minimal configuration.

See also

The Dockerfile can be used to specify several features of the image. Check the list of commands in the official documentation:

- *Dockerfile reference*: `https://docs.docker.com/reference/dockerfile/`

Also, you can have a look at the Docker CLI documentation on the following page:

- *Docker*: `https://docs.docker.com/reference/cli/docker/`

You can have a look at the FastAPI documentation page dedicated to the integration with Docker at this link:

- *FastAPI in Containers - Docker*: `https://fastapi.tiangolo.com/deployment/docker/`

Running the server across multiple workers

In high-traffic environments, running a FastAPI application with a single worker may not be sufficient to handle all incoming requests efficiently. To improve performance and ensure better resource use, you can run your FastAPI instance across multiple workers. This can be achieved using tools such as **Gunicorn**.

In this recipe, we will explore how to run a FastAPI application with multiple workers using Gunicorn in a Docker container, and we will also discuss Uvicorn's ability to handle multiple workers along with its limitations.

Getting ready

The `gunicorn` package is not Windows compatible. To ensure operating system operability, we will run our `Live Application` in a Docker container.

The recipe will be based on the project created in the previous recipe, *Running FastAPI applications in Docker containers*.

How to do it...

FastAPI with multiple workers runs multiple copies of the app on different CPU processes.

To see this better, let's make the endpoint show the **process ID** (**PID**) number of the process. In `main.py`, add these lines:

```
import logging
from os import getpid
# rest of the module
```

```
logger = logging.getLogger("uvicorn")
# rest of the module

@app.get("/")
def read_root():
    logger.info(f"Processd by worker {getpid()}")
    return {"Hello": "World"}
```

Let's add the gunicorn dependency in the requirements.txt file as follows:

```
fastapi
gunicorn
```

We will use gunicorn instead of uvicorn to run the server.

If you are on Linux or macOS, you simply install gunicorn in your environment like this:

```
$ pip install gunicorn
```

Then, run the server with four workers with the following command:

```
$ gunicorn app.main:app --workers 4 \
--worker-class uvicorn.workers.UvicornWorker
```

If you are on Windows, we will use Docker. In the Dockerfile.dev file, add the new CMD instruction below the existing one, which will be ignored:

```
CMD ["gunicorn",\
    "app.main:app",\
    "--bind", "0.0.0.0:80",\
    "--workers", "4",\
    "--worker-class",\
    "uvicorn.workers.UvicornWorker",\
    "--log-level", "debug"]
```

Then, build the Docker image with the following:

```
$ docker build -t live-application-gunicorn \
-f Dockerfile.dev .
```

Next, run the container from the image:

```
$ docker run -p 8000:80 -i live-application-gunicorn
```

The -i parameter allows you to run the container in interactive mode to see the logs.

After the server is running, open the browser on `http://localhost:8000/docs` and use the interactive documentation to make calls. On the terminal output, you will notice different PIDs that vary for each call.

This shows that Gunicorn can distribute the load among different processes, and you can take advantage of multiple CPU cores.

You have learned how to run a FastAPI app with Gunicorn and multiple workers, which can improve the performance and scalability of your web service. You can experiment with different settings and options to find the optimal configuration for your needs.

> **Important note**
>
> You can run multiple workers with Uvicorn as well. However, Uvicorn's worker process management is not as advanced as Gunicorn's at the moment.

There's more...

One of the benefits of running Gunicorn with multiple workers is that it can handle more concurrent requests and improve the performance and availability of the web application. However, there are also some challenges and trade-offs that come with this approach.

For example, when using multiple workers, each worker process has its own memory space and cannot share data with other workers. This means that any stateful components of the application, such as caches or sessions, need to be stored in a centralized or distributed service, such as Redis or Memcached. Moreover, multiple workers may increase resource consumption and the risk of contention on the server machine, especially if the application is CPU-intensive or input/output-bound. Therefore, it is important to choose the optimal number of workers based on the characteristics of the application and the available resources.

A common heuristic is to use the formula *workers = (2 x cores) + 1*, where *cores* means the number of CPU cores on the server. However, this may not be suitable for all scenarios and may require some experimentation and fine-tuning.

See also

You can discover more about Gunicorn in the official documentation at this link:

- *gunicorn:* `https://gunicorn.org/`

Also, you can have a look at the page in the FastAPI documentation dedicated to server workers:

- *Server Workers – Gunicorn with Uvicorn:* `https://fastapi.tiangolo.com/deployment/server-workers/`

Deploying your FastAPI application on the cloud

Deploying your FastAPI application on the cloud is an essential step to make it accessible to users worldwide. In this recipe, we will demonstrate how to deploy a FastAPI application on Railway.

Railway is a versatile and user-friendly platform that enables developers to deploy, manage, and scale their applications with ease. By the end of the recipe, you will have a FastAPI application running on Railway, ready to serve users on the internet.

Getting started

Before we begin, ensure that you have already set up an application, as we will be deploying it on the cloud. The recipe will be applied to our `Live Application`, the basic application created in the *Running the server with the FastAPI CLI* recipe.

Also, put the project folder on GitHub, since it will be used as a reference for the deployment.

You will also need to set up an account at `https://railway.app`. The creation is straightforward, and you can use your GitHub account as well. When you sign up, you will receive a $5 credit, which is more than enough to cover the recipe.

How to do it...

We will demonstrate how to deploy the application on Railway through the following steps:

1. Create the configuration file.
2. Connect the Git repository.
3. Configure the deployment.

   Although we will demonstrate it specifically for Railway, these steps are also common for other cloud services.

Creating the configuration file

Every deployment tool requires a configuration file that contains specifications for the deployment. To deploy on Railway, under our project root folder, let's create a file called `Procfile`. The file content will be as follows:

```
web: fastapi run --port $PORT
```

Remember to push the file to the GitHub repository hosting your project to be visible to Railway.

Connecting the Git repository

Once the configuration file is set up, log in to Railway (`https://railway.app/login`) with your account and you will be redirected to your dashboard (`https://railway.app/dashboard`).

Then, click on the **+ New Project** button at the top right of the screen. Once on the new page, choose the **Deploy from Github repo** option and select the repository that hosts your project. If you forked the `FastAPI-Cookbook` repository (`https://github.com/PacktPublishing/FastAPI-Cookbook`), you can select it.

Then select **Deploy now** and wait for the deployment to set up. It will automatically create a new project with a fictional name. Mine, for example, is `profound-enchantment`.

Once finished, the *deployment* icon will appear on the project dashboard. By default, the deployment takes the name of the chosen GitHub repository. In my case, it's `FastAPI-Cookbook`.

Configuring the deployment

When you click on the *deployment* icon, you can see a warning indicating that the deployment has failed. To resolve this, we need to add some parameters.

Click on the *deployment* icon, which will open a window on the left. Then, click on the **Settings** tab. This will display a list of configurations with sections such as **Source**, **Networking**, **Build**, and **Deploy**.

Begin with the **Source** section. If you've chosen the project from the `FastAPI-Cookbook` repository or if your project's root directory is not the repository root, click on **Add Root Directory** under the **Source** repository specification and enter the path.

For the `FastAPI-Cookbook` repository, the path will be `/Chapter12/live_application`. After adding the path, click on the *save* icon.

Leave the branch selected as **main**.

Moving on to the **Networking** section, click on the **Generate Domain** button under the **Public Network** subsection. This will create a unique domain for exposing your application. Mine is `fastapi-cookbook-production.up.railway.app`. You will have a slightly different domain.

Leave the remaining settings as they are.

At the top left of the screen, you will see a text bar with the text **Apply 2 changes** with a **Deploy** button. Click on it to apply the modification we have done.

After the deployment process is complete, your application will begin to handle live web traffic. The public address is defined in the **Networking** section of the **Settings** panel.

Open the address in a new browser tab, and check the response. You should see the implemented response:

```
{
    "Hello": "World"
}
```

In your web browser's address bar, you can see a *lock* icon, which indicates that the connection is secure and has a certificate. Usually, when you expose your service to the web, the hosting platform provides you with certificates.

You have just deployed your FastAPI application to be accessible on the World Wide Web. Now, users from all over the world can access your service.

There's more...

To deploy your service, Railway creates an image and then a container to run your service. You can specify a custom image with a Dockerfile and it will be automatically detected.

See also

You can discover more about Railway services on the official documentation website:

- *Railway Docs:* https://docs.railway.app/

You can check the official FastAPI template used for Railway at this link:

- *FastAPI Example:* https://github.com/railwayapp-templates/fastapi

FastAPI is one of the fastest-growing production applications, especially on the major public cloud service providers. That's why you can find extensive documentation on how to use it:

For **Google Cloud Platform (GCP)**, you can follow the article at the link:

- *Deploying FastAPI app with Google Cloud Run* article at the following link: https://dev. to/0xnari/deploying-fastapi-app-with-google-cloud-run-13f3

For **Amazon Web Services (AWS)**, check this Medium article:

- *Deploy FastAPI on AWS EC2:* https://medium.com/@shreyash966977/deploy-fastapi-on-aws-ec2-quick-and-easy-steps-954d4a1e4742

For Microsoft Azure, you can check the official documentation page:

- *Using FastAPI Framework with Azure Functions:* https://learn.microsoft.com/en-us/samples/azure-samples/fastapi-on-azure-functions/fastapi-on-azure-functions/

On the FastAPI website, you can check other examples for other cloud providers at the following link:

- *Deploy FastAPI on Cloud Providers*: `https://fastapi.tiangolo.com/deployment/cloud/`

A useful tool is the Porter platform, which allows you to deploy your applications on different cloud services such as AWS, GCP, and Azure from one centralized platform. Have a look at this link:

- *Deploy a FastAPI app:* `https://docs.porter.run/guides/fastapi/deploy-fastapi`

Shipping FastAPI applications with Hatch

Packaging and shipping a FastAPI application as a distributable package are essential for deploying and sharing your application efficiently.

Hatch is a modern Python project management tool that simplifies the packaging, versioning, and distribution process. In this recipe, we'll explore how to use Hatch to build and ship a package containing a FastAPI application. This will ensure that your application is portable, easy to install, and maintainable, making it easier to deploy and share with others.

Getting ready

Hatch facilitates the use of multiple virtual environments for our project. It uses the `venv` package under the hood.

To run the recipe, you need to install Hatch on your local machine. The installation process may vary depending on your operating system. Detailed instructions can be found on the official documentation page: `https://hatch.pypa.io/1.9/install/`.

Once the installation is complete, verify that it has been correctly installed by running the following from the command-line terminal:

```
$ hatch --version
```

You should have the version printed on the output like this:

```
Hatch, version 1.11.1
```

Make sure that you installed a version higher than `1.11.1`. We can then start creating our package.

How to do it...

We divide the process of shipping our FastAPI package into five steps:

1. Initialize the project.
2. Install dependencies.
3. Create the app.
4. Build the distribution.
5. Test the package.

Let's start building our package.

Initializing the project

We start by creating our project by bootstrapping the structure. Let's call our application FCA, which stands for **FastAPI Cookbook Application**. Let's bootstrap our project by running the following command:

```
$ hatch new "FCA Server"
```

The command will create a project bootstrap under the fca-server folder as follows:

```
fca-server
├──src
│   └── fca_server
│       ├── __about__.py
│       └── __init__.py
├──tests
│   └── __init__.py
├──LICENSE.txt
├──README.md
└──pyproject.tomt
```

We can then directly use a virtual environment by entering the fca-server directory and running the following:

```
$ hatch shell
```

The command will automatically create a default virtual environment and activate it. You will see your command-line terminal with a prepend value, (fca-server), like so:

```
(fca-server) path/to/fca-server $
```

Verify that the environment is correctly activated by checking the Python executable. You do it by running the following:

```
$ python -c "import sys; print(sys.executable)"
```

The executable should come from the virtual environment called `fca-server`, which will present a path such as `<virtual environment locations>\fca-server\Scripts\python`.

This will give you information on the virtual environment that you can also provide to your **integrated development environment** (**IDE**) to work with the code.

You can exit from the shell by typing `exit` in the terminal. Also, you can run commands in the virtual environment without spawning the shell. For example, you can check the Python executable of the default environment by running the following:

```
$ hatch run python -c "import sys; print(sys.executable)"
```

We can now proceed to install the package dependencies in our environment.

Installing dependencies

Now that you have created a virtual environment, let's add the `fastapi` dependency to our project. We can do it by modifying the `pyproject.toml` file. Add it in the `dependencies` field under the `[project]` section like so:

```
[project]
...
dependencies = [
  "fastapi"
]
```

Next time you spawn a shell, the dependencies will synchronized and the `fastapi` package will be installed.

Let's see, for example, whether the `fastapi` command works by running the following:

```
$ hatch run fastapi --help
```

If you see the help documentation of the command, the dependency has been added correctly.

Creating the app

Now that we have the environment with the `fastapi` package installed, we can develop our application.

Let's create the `main.py` module under the `src/fca_server` folder and initialize the `APIRouter` object with one endpoint like this:

```python
from fastapi import APIRouter


app = APIRouter()


@app.get("/")
def read_root():
    return {
        "message":
        "Welcome to the FastAPI Cookbook Application!"
    }
```

Then, let's import the router into the `src/fca_server.__init__.py` file as follows:

```python
from fca_server.main import router
```

This will allow us to directly import the router from the `fca_server` package from an external project.

Building the distribution

Now that we have finalized the package, let's leverage Hatch to build the package distribution.

We will generate the package in the form of a `.tar.gz` file by running the following:

```
$ hatch build -t sdist ../dist
```

It will generate the `fca_server-0.0.1.tar.gz` file placed outside of the project in a `dist` folder. We will then use the file in an external project.

Testing the package

Next, we will make a different project that uses the `fca_server` package we made.

Create an `import-fca-server` folder outside of the `fca-server` folder for the package and use it as the project root folder.

In the folder, make a local virtual environment with `venv` by running the following:

```
$ python -m venv .venv
```

Activate the environment. On Linux or macOS, type the following:

```
$ source .venv/Scripts/activate
```

On Windows, type this instead:

```
$ .venv\Scripts\activate
```

Install the `fca_server` package with `pip`:

```
$ pip install ..\dist\fca_server-0.0.1.tar.gz
```

Use the path where the `fca_server-0.0.1.tar.gz` file is.

Now, try to import the package.

Make a `main.py` file and import the router from the `fca_server` package:

```
from fastapi import FastAPI
from fca_server import router

app = FastAPI(
    title="Import FCA Server Application"
)


app.include_router(router)
```

Run the server from the command line:

```
$ fastapi run
```

Go to the interactive documentation at `http://localhost:8000/docs` and see the endpoint in the external package. You have just created a custom package and imported it into another project.

You have learned how to use Hatch to create and manage your Python projects with ease. This is a powerful tool that can save you time and effort and help you write better code. Now, you can experiment with different options and features of Hatch and see what else you can do with it.

There's more...

Hatch is a versatile packaging system for Python that allows you to create scripts and multiple environments for your projects.

With Hatch, you can also customize the location of the virtual environment files, such as whether you want them to be centralized or in the project folder. You can specify this option in the `config.toml` file, which contains the configuration settings for Hatch.

To find the location of the `config.toml` file, you can run the following command in your terminal:

```
$ hatch config find
```

Hatch also lets you create the build of your package in a wheel format, which is a binary distribution format that is more efficient and compatible than the traditional source distribution.

Moreover, you can publish your package directly to the **Python Package Index (PyPI)**, where other users can find and install it. Hatch makes it easy to share your code with the world.

See also

You can find more information about Hatch in the official documentation at

- *Hatch*: `https://hatch.pypa.io/latest/`

We learned how to create a project bootstrap, but with Hatch, you can also initialize an existing project. Check out the documentation page:

- *Existing project:* `https://hatch.pypa.io/1.9/intro/#existing-project`

One of the greatest advantages of using Hatch is the flexibility of running the project for several virtual environments. Check more on the documentation page:

- *Environments:* `https://hatch.pypa.io/1.9/environment/`

The `pyproject.toml` file is a configuration file for Python projects, introduced in `PEP 518` (`https://peps.python.org/pep-0518/`). It aims to standardize and simplify the configuration of Python projects by providing a single place to specify build system requirements and other project metadata. It is used by other build tools. You can have a look at the Python Package User Guide page at the following link:

- *Writing your pyproject.toml:* `https://packaging.python.org/en/latest/guides/writing-pyproject-toml/`

You can see more on how to manage Python dependencies on this page:

- *Dependency configuration:* `https://hatch.pypa.io/dev/config/dependency/`

Index

REST APIs 249

RESTful API

testing 59-64

RESTful APIs 51

documenting, with Redoc 74, 75

documenting, with Swagger UI 74, 75

RESTful Endpoints

creating 57-59

Retrieval-Augmented Generation
(RAG) 267

role-based access control (RBAC) 77, 188

setting up 88-93

S

scikit-learn documentation

reference link 262

Secure Sockets Layer/Transport Layer
Security (SSL/TLS) certificate 305

semantic versioning 68

sensitive data

exposing, from NoSQL databases 183-188

securing, best practices 47-49

securing, in SQL databases 156-160

serialization

working with 36, 37

serialization concept 37, 38

server

running, across multiple workers 311-313

running, with FastAPI CLI 303, 304

session cookies

handling 105-108

Slowapi features

URL 222

software as a service (SaaS) 77

SQLAlchemy 25, 135

CRUD operations with 28, 29

database connection, initializing 139

mapping object classes, creating 137-139

reference link 28, 32

setting up 136

SQL database

connection, establishing 27, 28

setting up 25-27

SQL databases

sensitive data, securing in 156-160

SQLite database 79, 136

SQLite In-Memory Database Configuration

reference link 120

SQL queries, for performance

data, minimizing to fetch 155, 156

join statement, using 154, 155

N+1 queries, avoiding 153, 154

optimizing 152, 156

Stack Overflow

reference link 211

Starlette library 230

Structured Query Language (SQL) 23

Swagger UI

used, for documenting API 74, 75

using 13

system message 264

T

Task Manager API

complex queries and filter, handling 64-66

securing, with OAuth2 69-73

versioning implementation 66-68

techniques, for optimizing
FastAPI performances

asynchronous programming 218

caching 218

Uvicorn workers, scaling 218

testing environment

setting up 110-112

www.packtpub.com

Subscribe to our online digital library for full access to over 7,000 books and videos, as well as industry leading tools to help you plan your personal development and advance your career. For more information, please visit our website.

Why subscribe?

- Spend less time learning and more time coding with practical eBooks and Videos from over 4,000 industry professionals

- Improve your learning with Skill Plans built especially for you

- Get a free eBook or video every month

- Fully searchable for easy access to vital information

- Copy and paste, print, and bookmark content

Did you know that Packt offers eBook versions of every book published, with PDF and ePub files available? You can upgrade to the eBook version at packtpub.com and as a print book customer, you are entitled to a discount on the eBook copy. Get in touch with us at customercare@packtpub.com for more details.

At www.packtpub.com, you can also read a collection of free technical articles, sign up for a range of free newsletters, and receive exclusive discounts and offers on Packt books and eBooks.

Other Books You May Enjoy

If you enjoyed this book, you may be interested in these other books by Packt:

Hands-On Microservices with Django

Tieme Woldman

ISBN: 978-1-83546-852-4

- Understand the architecture of microservices and how Django implements it
- Build microservices that leverage community-standard components such as Celery, RabbitMQ, and Redis
- Test microservices and deploy them with Docker
- Enhance the security of your microservices for production readiness
- Boost microservice performance through caching
- Implement best practices to design and deploy high-performing microservices

Node.js for Beginners

Ulises Gascón

ISBN: 978-1-80324-517-1

- Build solid and secure Node.js applications from scratch

- Discover how to consume and publish npm packages effectively

- Master patterns for refactoring and evolving your applications over time

- Gain a deep understanding of essential web development principles, including HTTP, RESTful API design, JWT, authentication, authorization, and error handling

- Implement robust testing strategies to enhance the quality and reliability of your applications

- Deploy your Node.js applications to production environments using Docker and PM2

Packt is searching for authors like you

If you're interested in becoming an author for Packt, please visit `authors.packtpub.com` and apply today. We have worked with thousands of developers and tech professionals, just like you, to help them share their insight with the global tech community. You can make a general application, apply for a specific hot topic that we are recruiting an author for, or submit your own idea.

Share Your Thoughts

Now you've finished *FastAPI Cookbook*, we'd love to hear your thoughts! Scan the QR code below to go straight to the Amazon review page for this book and share your feedback or leave a review on the site that you purchased it from.

`https://packt.link/r/1-805-12785-3`

Your review is important to us and the tech community and will help us make sure we're delivering excellent quality content.

Download a free PDF copy of this book

Thanks for purchasing this book!

Do you like to read on the go but are unable to carry your print books everywhere?

Is your eBook purchase not compatible with the device of your choice?

Don't worry, now with every Packt book you get a DRM-free PDF version of that book at no cost.

Read anywhere, any place, on any device. Search, copy, and paste code from your favorite technical books directly into your application.

The perks don't stop there, you can get exclusive access to discounts, newsletters, and great free content in your inbox daily

Follow these simple steps to get the benefits:

1. Scan the QR code or visit the link below

https://packt.link/free-ebook/978-1-80512-785-7

2. Submit your proof of purchase
3. That's it! We'll send your free PDF and other benefits to your email directly

www.ingramcontent.com/pod-product-compliance
Lightning Source LLC
LaVergne TN
LVHW081514050326
832903LV00025B/1487